INTRODUCTION TO GEOMETRIC DIMENSIONING AND TOLERANCING

by LOWELL W. FOSTER

National Tooling &
Machining Association

FOREWORD

This text is designed to provide students and others with a basic knowledge of geometric dimensioning and tolerancing. Because of its fundamental approach and specialized coverage, this text is a good primer on this increasingly important industrial topic.

"Introduction to Geometric Dimensioning and Tolerancing" will give the reader the personal confidence and basic skills needed to use and understand this topic. The reader will finish with a good foundation on which to acquire further knowledge.

For further study in this field, I recommend the more advanced text entitled "Modern Geometric Dimensioning and Tolerancing," also authored by Lowell Foster and published by the National Tooling and Machining Association.

I wish you success in learning and in the future use of this important communication tool.

Matthew B. Coffey
President
National Tooling and
    Machining Association

ABOUT THE AUTHOR

Lowell W. Foster

Lowell W. Foster is President and Director of Lowell W. Foster Associates, Inc., an engineering consulting firm in Minneapolis, Minnesota. He was with Honeywell, Inc. for thirty-one years in various technical and management capacities. His background includes tool and standards engineering, manufacturing, product design, and quality control, in addition to operating management and corporate staff responsibilities.

Mr. Foster is the author of thirty-six books and over 100 papers and presentations on metrication, standards, and production engineering. As an active lecturer and teacher, he has conducted over 1300 seminars on geometrics, effective design for productive manufacturing and product definition for over 10,000 companies, schools, professional societies, government agencies, and foreign countries.

He has been, or is, a member of the board of directors of the American National Standards Institute (ANSI), the American National Metric Council (ANMC), the International Standards Organization (ISO), the Standards Engineering Society (SES), the Society of Manufacturing Engineers (SME), the National Society of Professional Engineers (NSPE), and the American Institute for Design & Drafting (AIDD). He is a Certified Standardization Engineer, Certified Manufacturing Engineer, Registered Professional Engineer, Certified Advanced Metrication Specialist, and Certified Teacher in the State of Minnesota.

Mr. Foster attended the United States Coast Guard Academy and the University of Minnesota.

ACKNOWLEDGEMENTS

I wish to acknowledge the support of my wife, Marion, who acted as my secretary, typist, confidante and critic in this endeavor, for her help and encouragement, and my son, John, for his technical assistance and support. I also wish to thank my many friends and associates in standards work around the world and the companies, military, schools, and societies with whom it has been my pleasure to work. All have had an influence in my professional life and the efforts on this subject. My thanks also go to the National Tooling and Machining Association for its confidence and trust in supporting the development of this text.

Credit is also gratefully given and acknowledgement made for certain references and definitions derived from ANSI Y14.5M-1982 published by the American Society of Mechanical Engineers (ASME), New York, New York.

This text is written in keeping with the principles of United States standard ANSI Y14.5M-1982. The text uses various methods of explaining the subject matter with a variety of examples followed by Quiz-Exercises to reinforce your knowledge.

It is assumed that the reader has a prior basic knowledge of the standard conventions of drawing practice such as views, sections, assemblies, size dimensions and tolerances, etc. Such knowledge is a prerequisite for extending one's ability into learning and using the geometric dimensioning and tolerancing system.

This text is written using the customary inch based system of measurement. However, the principles and symbology could by applied equally using the metric based system of measurement as is shown at various places in the text.

Lowell W. Foster

Reference Documents

1. American National Standard ANSI Y14.5M-1982, *Dimensioning and Tolerancing*. American Society of Mechanical Engineers, New York, New York.

2. *Modern Geometric Dimensioning and Tolerancing with Workbook Section* and *Answerbook and Instructor's Guide for Modern Geometric Dimensioning and Tolerancing*. National Tooling and Machining Association, Ft. Washington, Maryland.

3. *Geometrics II, The Application of Geometric Tolerancing Techniques*, Addison-Wesley Publishing Co., Reading, Massachusetts.

4. *Tolerances of Form and of Position, Part I*, International Standards Organization (ISO), Standard 1101-1983, ISO, Geneva, Switzerland.

TABLE OF CONTENTS

1

INTRODUCTION

The manufacture of a product requires a plan and communication. That is, if a product such as an automobile, a computer, a switch, or a control is to be manufactured, it first must be conceived and planned. The plan is commonly referred to as the engineering design or drawing. Whether the drawing is prepared with manual drafting tools or equivalent computer aided tools, a picture of the product results. The picture or drawing defines the product to be manufactured. If the drawing clearly conveys the information required, the results should be satisfactory. Good communication has occurred. If the drawing is not clear, there is a breakdown in communication from design to manufacture and then to inspection of the final result.

Drafting practices, based upon well-established principles and standards, provide the basic graphic language for drawing preparation. The drawing views of the products and those of the individual component piece-part, as detailed, provide the graphic language. Adding dimensions and tolerances to each part drawing provides its magnitude (how big) and its precision (how close to perfect must it be).

It would seem that a drawing, prepared with these common standard drafting practices used properly, should produce a satisfactory result. This, unfortunately, is not so on many designs. The part ''geometry,'' or its 3-dimensional (3-D) shape, is not controlled by conventional dimensions and tolerances. Size dimensions (how big) and their tolerances control only themselves and not relationships to other part features. A feature is a physical portion of a piece-part (a hole, pin, surface, etc.). Geometric dimensioning and tolerancing provides the added necessary ingredients to complete the part description. This controls such relationships and also any other required refinements.

Geometric dimensioning and tolerancing adds the ability to further indicate the form (how flat, how circular, etc.) and also to relate features. It builds upon established engineering drawing practices so that the part can be better described along with its function or relationship in the product assembly (how perpendicular, how parallel, location to other features, etc.). Thus, such a drawing conveys a clearer picture to manufacturing and quality or inspection. The necessary tools, equipment, processes and operations can be selected and used in keeping with the understood requirements. Geometric dimensioning and tolerancing (geo-metrics) is really a universal language for communications. It also provides new techniques and methods which vastly improve design efficiency. This facilitates the production cycle and often increases the available tolerances.

Geometric dimensioning and tolerancing principles are based upon standards as established over years of experience and development. American National Standard, ANSI Y14.5M-1982 *Dimensioning and Tolerancing,* is the current basis of authority from which the state-of-the-art in the United States is derived. These practices are also compatible with international standards such as ISO 1101, *Technical Drawings-Geometrical Tolerances of Form, Orientation, Profile, Runout and Location.* Thus, even international interchange of drawings and documentation, etc., is assisted by use of these principles and standards.

This text explains the fundamentals of geometric dimensioning and tolerancing. Its aim is to provide the reader with a basic knowledge of the system, its application and meaning. Further knowledge can be derived from experience and in-depth study.

PICTORIAL INTRODUCTION TO GEOMETRIC DIMENSIONING AND TOLERANCING

The introductory paragraphs explain the reasons and principles behind the use of geometric dimensioning and tolerancing. They introduce the idea that the "geometry" (relationship) of the concerned part features in an assembly could be important but may not be adequately controlled by the size dimensions and tolerances. Thus, geometric tolerances are necessary to define the parts.

Using illustrations, let us extend the meaning of the preceding paragraphs:

The assembly of three parts is shown in Fig. 1-1. This is the manner in which the parts should assemble if properly made. What do we mean by properly made? Dimensions (the ideal) and tolerances (the amount of accepted deviation from the ideal) are needed to ensure proper manufacture and mating of the parts. They are necessary to control the part "geometry" beyond that which size can be expected to control.

In Fig. 1-2, the same assembly obviously appears to be a different situation. The flange mount part appears "bad" and will not assemble with its mating parts. However, even though the illustration greatly exaggerates the errors, the part could still meet all the size dimensions. It is a "good" part according to size controls. But, its geometry (relationships) is not controlled by size, therefore, the part is functionally bad.

How much its geometry can vary from perfect (Fig. 1-1) and remain acceptable to assembly (maximum conditions in Fig. 1-2) requires geometric tolerance controls. When geometric dimensioning requirements are added, and the parts manufactured to those requirements, it could vary in the manner shown in Fig. 1-2 but only within the stated acceptable limits (tolerances). This would guarantee assembly of the two parts. It is seen that geometric dimensioning and tolerancing is an essential tool to specify and ensure assembly of such parts.

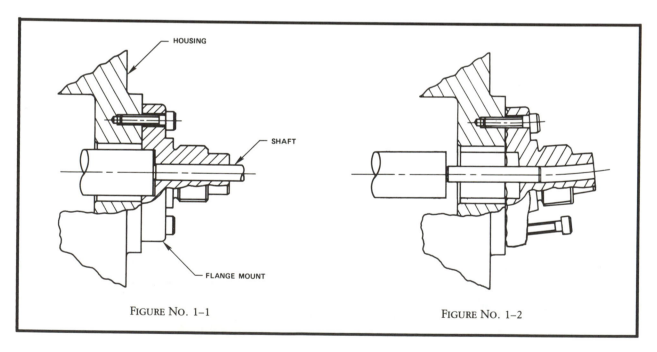

FIGURE NO. 1–1   FIGURE NO. 1–2

COMPARISON BETWEEN COORDINATE DIMENSIONING AND TOLERANCING AND GEOMETRIC DIMENSIONING AND TOLERANCING

The discussion in this introduction is not designed to teach you geometric tolerancing application. Its purpose is to introduce the reasoning or the "why," of geometric dimensioning and tolerancing. The following text sections will address the details of selection and application step-by-step. At this point, we introduce the "idea" of geometric tolerancing before the details of "how" are addressed. The Flange Mount shown in Figures 1-1 and 1-2 is used as our example.

Figures 1-3 and 1-4 compare a drawing prepared using only the coordinate dimensioning and tolerancing method (Fig. 1-3) and one where the necessary geometric dimensioning and tolerancing has been added (Fig. 1-4). Although the meaning of the boxes and symbols in Fig. 1-4 are not expected to be known at this time, note that these are typical examples of geometric dimensioning and tolerancing requirements and the general manner in which they are placed on the drawing. Adding such geometric dimensioning and tolerancing more clearly conveys the designer's intent and requires that the part be manufactured to these specifications. This ensures that the part will assemble according to plan.

Coordinate Tolerance Versus Geometric Tolerance

Coordinate dimensioning and tolerancing, as shown in Fig. 1-3, controls only "size" (how big) of features (holes, pins, etc.) when specified as illustrated with the $.315^{+.003}_{-.000}$ and $.980^{+.005}_{-.000}$ dimensions. From Fig. 1-1, it appears necessary that the .315 hole be somehow controlled in its squareness (perpendicularity) to the flange mount surface which contacts the housing surface in assembly. Otherwise, improper assembly relationships could occur between the .315 hole, the mounting surface of the flange and the shaft extending from the housing into the .315 hole. However, there is no assurance there will be a satisfactory assembly of these features as would result from the dimensional callouts shown in Fig. 1-3. The results would be left entirely to chance. If no mention is made of such a relationship, how does anyone know of such a requirement in manufacture or inspection of the $.315^{+.003}_{-.000}$ diameter hole? Will the hole be produced in a squareness adequately precise or not? Perfect parts cannot be produced. It is quite certain that different results will be found dependent upon each individual situation. The .315 hole must, of course, meet its size tolerance requirement, but could fail its assembly obligation as shown in Fig. 1-1. Note that in Fig. 1-4 the .315 hole is controlled with a perpendicularity requirement. A geometric tolerance has been added. A datum feature symbol has also been added to identify to which surface the $.315^{+.003}_{-.000}$ hole is to be related. A clear meaning for all to understand is now found on the drawing.

To continue with the comparison in more detail, note that the .315 hole and the counterbore $.980^{+.005}_{-.000}$ as specified in Fig. 1-3 are individual and separate size requirements. Again, what controls the relationship of the .315 and .980 features? Answer: Nothing, other than chance. Such factors as "good workmanship," "common sense," "probabilities" and "standards" usually exist to some extent behind the scenes to assist the chance of producing a part which will assemble properly. These are all important ingredients in a manufacturing situation and contribute to any success which results. However, unless such factors are specifically defined and applied with common understanding of their values and limits, the results are yet in question. Such is the case, we must assume, on a drawing where it is left to "explain itself." Ambiguity and unanswered questions invariably result. Note in Fig. 1-4 that the $.980^{+.005}_{-.000}$ counterbore is related to the $.315^{+.003}_{-.000}$ hole with a runout requirement (also note the datum B references). Again, a geometric tolerance has been added. A clear understanding of the feature relationships is now possible by reading the drawing.

PART DRAWING USING COORDINATE DIMENSIONING AND TOLERANCING

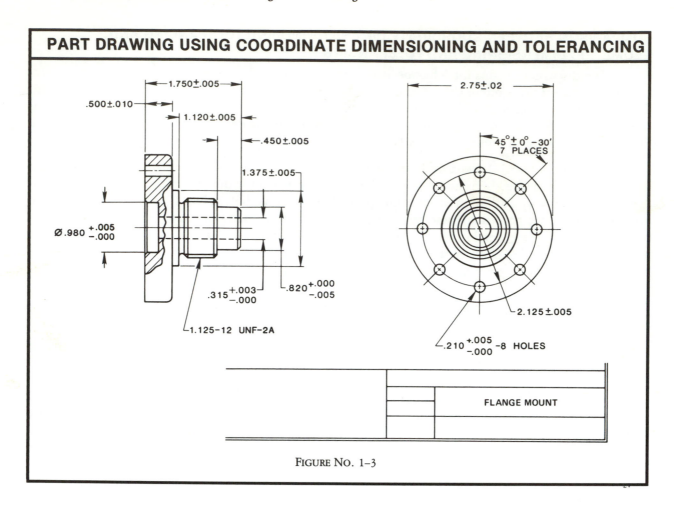

FIGURE NO. 1–3

Geometric Tolerance Provides Clear Communication

Note that with geometric tolerances added, communication of the design requirements are clear, manufacturing better understands what is required and inspection can uniformly determine if such requirements have been met. The drawing has clearly communicated its meaning to all parties concerned in a manner that vastly improves the technical efficiency of the design to manufacture cycle. Because of this efficiency, the cost-effectiveness of transmitting such information is an obvious economic advantage as well as a technical one.

Coordinate Tolerance Versus Geometric Tolerance For Location

In Fig. 1-3, the eight $.210^{+.005}_{-.000}$ holes are located by the 45 degree angles and the 2.125 bolt circle. The tolerances indicated for these dimensions ($\pm 0° - 30'$ and $\pm .005$) may be questioned immediately. How were they developed? What is the basis for their calculations? Where is the vertex of the 45 degree angle? Where is the center of 2.125 bolt circle? Are the holes' squareness controlled to something? Most of these questions are unanswerable from the information found on this drawing. Fig. 1-1 clearly shows the desired relationships in theory. But how are such requirements communicated specifically?

In Fig. 1-4, the callouts associated with the .210 holes, the boxed $\boxed{45°}$ and 2.125 dimensions, and the datum referenced features A and C answer all of the foregoing questions. Geometric tolerancing methods have been used.

PART DRAWING USING GEOMETRIC DIMENSIONING AND TOLERANCING

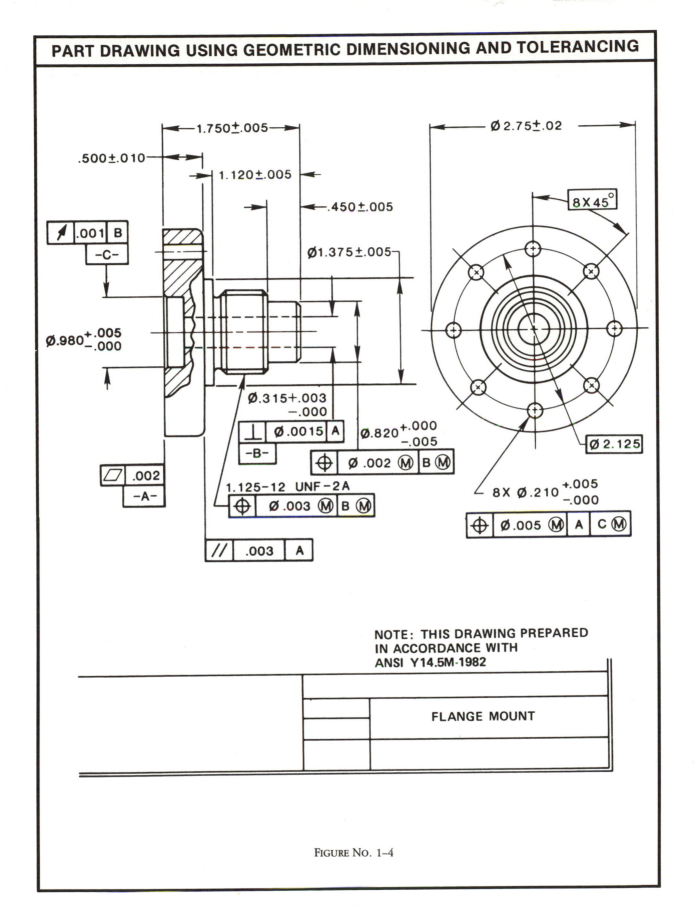

NOTE: THIS DRAWING PREPARED
IN ACCORDANCE WITH
ANSI Y14.5M-1982

FLANGE MOUNT

Figure No. 1–4

At this point we may not clearly understand the meaning of symbology. The purpose here is to provide some preliminary awareness of the general principles and motivating reasons for geometric tolerancing. Further study of the text will reveal all of the necessary details to acquire an introductory understanding of such application. You will, for example, learn the simple methods of calculating location tolerance for hole or pin location on mating parts similar to those shown in Fig. 1-1 and specified on the geometric toleranced example (Fig. 1-4). Likewise, detailed coverage will be given to other geometric tolerance controls (flatness, parallelism) and additional symbols and callouts as shown in Fig. 1-4.

Geometric Dimensioning And Tolerancing Captures Part Function

The geometric dimensioning and tolerancing system provides the ability for the design to convey requirements more clearly. The 3-dimensional relationships and special geometrical requirements represent the part function (what the part is to do). If the part drawing can capture the "function" of the part and its "relationships" to other features, there is much greater assurance of a satisfactory result. A valid way of describing geometric dimensioning and tolerancing is to say it is a graphic language which represents the necessary "function" and "relationship" of part features.

FUNCTION and RELATIONSHIP are the key words!

THE GEOMETRIC CHARACTERISTICS AND SYMBOLS

Geometric dimensioning and tolerancing is a graphic language using numbers and symbols. Symbols are used because they are of universal meaning and provide the language of the system. The symbols are established in the recognized national and international standards (ANSI Y14.5M-1982 and ISO 1101-1983) which provide the basis of authority for application. This text is based upon the United States standard principles found in ANSI Y14.5M-1982.

The geometric dimensioning and tolerancing system, using symbols, provides the ability to state requirements clearly. These requirements would otherwise be ambiguous or simply not be stated at all. Thus, the manufacture or inspection of such parts cannot be completed with any degree of assurance that acceptable parts will result. However, with the use of the "system" and symbolization, communication is vastly improved. In fact, a new ability to communicate information is possible.

Learning the Symbols

The symbols for the kinds of control which may be desired are shown on the opposite page. They are called Geometric Characteristics. Becoming familiar with these symbols at this time will begin the learning process and start the introduction of the technical details of the geometric dimensioning and tolerancing (or geometrics) system.

Note that some geometric characteristics "look like" the controls they represent. Many come from geometry books as well-established symbols familiar to many. Some symbols, such as profile and position, are a bit more broadly applied to varying situations and do not necessarily "look like" the control or shape of the feature concerned. However, these are unique to certain "kinds" of control and, thus, also assist the user in selecting the proper characteristic. Position tolerance, widely used to control holes and pins in location, "looks like" the hole or pin, does it not?

Study and learn the geometric characteristics and symbols as the tools or building blocks of the system. They will soon become natural to you in speaking the language of geometrics.

THE GEOMETRIC CHARACTERISTICS AND SYMBOLS

FORM TOLERANCES	▱ FLATNESS
	— STRAIGHTNESS
	○ CIRCULARITY (ROUNDNESS)
	⌭ CYLINDRICITY
ORIENTATION TOLERANCES	⊥ PERPENDICULARITY (SQUARENESS)
	∠ ANGULARITY
	// PARALLELISM
PROFILE TOLERANCES	⌒ PROFILE OF A LINE
	⌓ PROFILE OF A SURFACE
RUNOUT TOLERANCES	↗ CIRCULAR RUNOUT
	↗↗ TOTAL RUNOUT
LOCATION TOLERANCES	⊕ POSITION
	◎ CONCENTRICITY

OTHER RELATED SYMBOLS AND TERMS

Other symbols and the terms they represent are shown on the opposite page. It is not necessary that you learn them all yet at this point. Nevertheless, these symbols and terms should now be made familiar as a start. They will be reintroduced when and where they become necessary in following sections of this text. The symbols for BASIC, DATUM FEATURE SYMBOL, MAXIMUM MATERIAL CONDITION, DIAMETRICAL TOL ZONE OR FEATURE and FEATURE CONTROL FRAME are used in Fig. 1-4. Refer to the drawing and see if some of the reasoning and methods shown do not already start your learning of geometric tolerancing. Getting the ''idea'' and recognizing some of the reasoning and methods should be first. The mechanics of the system and detailed explanation can then follow successfully.

THE FEATURE CONTROL FRAME

The Feature Control Frame is the pictorial ''note'' or ''message'' being sent or received regarding the details of a particular requirement. Begin to understand its content, meaning and symbolic makeup. The summary at right acquaints us with the composition of the Feature Control Frame.

The Feature Control Frame may be for a relatively simple callout, such as the flatness requirement (opposite page-lower left), or the more involved type such as position (lower right). Similar callouts have been used in Fig. 1-4. Without concern for the detailed meaning of the symbols at this time, try to become acquainted with the general purpose and appearance of the Feature Control Frame. It is used throughout the system and you should recognize its importance and prominence early in your learning efforts.

Do not yet concern yourself with the modifiers, (Ⓜ, Ⓢ, Ⓛ), their meaning, or application. Note only as shown at right and Fig. 1-4, where they appear in the Feature Control Frame. Discussion on modifiers will occur later in the text. See also Supplemental Information, page 177-179.

OTHER RELATED SYMBOLS AND TERMS

.605	BASIC, OR EXACT, DIMENSION
−A−	DATUM FEATURE SYMBOL
Ⓜ	MAXIMUM MATERIAL CONDITION
Ⓢ	REGARDLESS OF FEATURE SIZE
Ⓛ	LEAST MATERIAL CONDITION
Ⓟ	PROJECTED TOLERANCE ZONE
∅	DIAMETRICAL (CYLINDRICAL) TOL ZONE OR FEATURE
⊕ ∅ .005 Ⓜ A	FEATURE CONTROL FRAME
A1	DATUM TARGET SYMBOL

THE FEATURE CONTROL FRAME

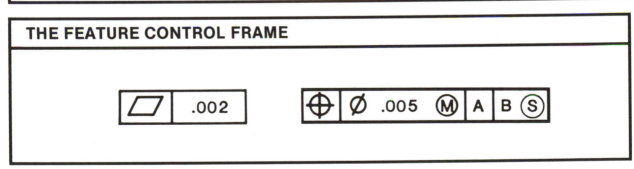

The feature control frame comprises the pictorial "note" which includes:

    a. Kind of control (geometric characteristic),

    b. The geometric tolerance,

    c. Any modifiers (ⓂⓈⓁ),

    d. Datum references, and any datum reference modifiers (ⓂⓈⓁ).

GEOMETRIC DIMENSIONING AND TOLERANCING USING METRIC SYSTEM

Geometric dimensioning and tolerancing can be applied using either the customary inch system or the metric system. It makes no difference. Since each system uses a linear value described by a number, either method is acceptable to geometric dimensioning and tolerancing practices.

Metric Standards

The United States standard ANSI Y14.5M-1982 and the ISO 1101-1983 standards are both written in the metric language. The US standard, however, permits continued use of the inch system as an option. In the transitional increase of metric terms in the US, industry, government and business must be flexible to be able to adapt to either methods. This text is written primarily in the inch system, but any example or discussion could equally have been presented in the metric system; some are shown.

Metric System Example

Fig. 1-5 uses the metric system as a basis for the drawing and the geometric tolerancing requirements. Note that the metric system methods of applying the coordinate dimensions (the ± 's) are according to the established standards (ANSI Y14.5M-1982). The geometric tolerance values are also metric instead of inch. A comparison of the metric drawing at right with the inch drawing in Fig. 1-4 is helpful in showing the similarities and the differences of the two methods. The values in the metric version are not necessarily direct translations of the inch values (Fig. 1-4). Figure 1-5 uses selected metric values as based upon preferred metric numbers.

Metric or Inch Use

As is seen from Fig. 1-5 and the similar inch based drawing Fig. 1-4, the geometric dimensioning and tolerancing remains identical. When a metric based drawing is prepared for possible world-wide use, additional drawing format considerations may be necessary. For example, it is indicated that the drawing is prepared in the Metric System and the projection used (first angle or third angle) is designated.

Whether the inch or metric system is used, all examples, principles and methods shown in this text are equally valid.

PART DRAWING USING GEOMETRIC DIMENSIONING AND TOLERANCING USING METRIC SYSTEM

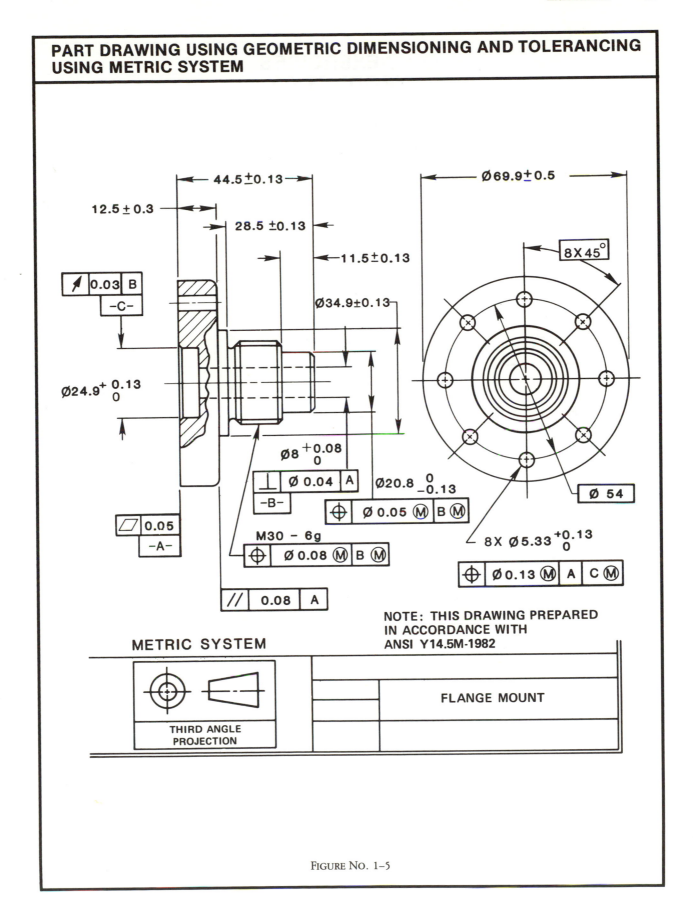

METRIC SYSTEM

THIRD ANGLE PROJECTION

FLANGE MOUNT

NOTE: THIS DRAWING PREPARED IN ACCORDANCE WITH ANSI Y14.5M-1982

Figure No. 1–5

1
QUIZ-EXERCISES
INTRODUCTION

The following questions are relative to the material in this chapter. Read the question and answer to the best of your ability. The answers can be found in the companion manual *Answer Book and Instructor's Guide for Introduction to Geometric Dimensioning and Tolerancing*.

Consult your instructor if you have any questions.

1. Geometric Dimensioning and Tolerancing provides numerous advantages. List at least four of them.

   a. _____

   b. _____

   c. _____

   d. _____

   e. _____

   f. _____

   g. _____

2. In determining geometric dimensioning and tolerancing requirements and application, the two key words to remember for guidance are _____ and _____.

3. Which two of the following considerations for applying geometric tolerancing are usually most important in determining when the system should be used on a given application? (Check two.)

   _____ Follow national standard practices.

   _____ Where size of features do not provide adequate control.

   _____ Where feature relationships are to be specified.

4. For delineating these requirements on a drawing, the method that provides best uniformity and efficiency and is recommended by the National Standard ANSI Y14.5, is: (Check one.)

   _____ symbolically

   _____ by note

5. Here are the thirteen varieties of geometric characteristics. Place the correct symbol designation beside each one.

   Flatness   Circular runout

   Straightness   Total runout

   Angularity   Profile of a surface

   Perpendicularity   Profile of a line

   Parallelism   Position

   Circularity   Concentricity

   Cylindricity

6. List the four elements of geometric control that may be used in making up a complete Feature Control Frame.

   a. _____

   b. _____

   c. _____

   d. _____

7. Fill in the Feature Control Frame below so that it states "the feature surface is to be flat within .005."

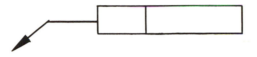

—————— **NOTES** ——————

—————————————————————————

—————————————————————————

—————————————————————————

—————————————————————————

—————————————————————————

—————————————————————————

2
MEANING OF A SIZE DIMENSION AND TOLERANCE

The first type of dimension and tolerance found in geometric dimensioning and tolerancing is that of a size feature such as hole, pin, width, etc.

The introductory discussions emphasized that size is important and that such tolerances must be met in production. It was also emphasized that feature size tolerances control only its "size" and nothing more.

Now we will discuss more about a size feature and its control and how it must always be in consideration when determining if, when and how geometric dimensioning and tolerancing is necessary.

DEFINITIONS

"Feature" is the general term applied to a physical portion of a part such as a surface, hole, pin, slot, tab, etc.

"Feature of size" is one cylindrical or spherical surface, or a set of two plane parallel surfaces, each of which is associated with a size dimension.

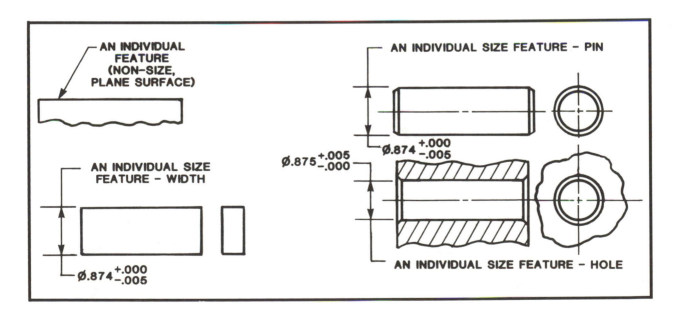

The meaning of the word feature is both general (as defined above) and specific when certain geometric controls are applied to features. In the latter case, when applied with geometric tolerances, the meaning is automatically clarified when the selected geometric characteristic is specified. (Flatness is applied only to a non-size plane surface feature. Position is applied only to size features.)

15

LIMITS OF SIZE

Where the geometric dimensioning and tolerancing standard is established as the authority, the size dimensions and tolerances also control the form of that individual size feature as well. That is, geometric tolerance of a composite variety is automatically included whenever a size dimension and tolerance is placed on the part drawing. This applies only, however, to the individual feature and not to relationships of features.

This principle of "size" controlling "form" of the individual feature has been established in USA and world standards for many years. It provides the beginning basis (worst case, extreme limit) for the control of feature size and geometry from which we expand the controls (geometrics) as necessary. This exacting beginning basis (a theoretically perfect size-extreme limit) is necessary as a ground rule. Otherwise, what could size mean? Answer: Almost anything. This, of course, is not an acceptable answer to modern design and manufacture.

The principle previously discussed is so important as a design basis that it is called Rule #1—The Limits of Size Rule in the USA standard.

RULE #1—THE LIMITS OF SIZE RULE

Individual Features of Size

Where only a tolerance of size is specified, the limits of size of the individual feature prescribe the extent to which variations in its geometric form as well as size are allowed.

Variations of Form

The form of an individual feature is controlled by its limits of size to the extent prescribed in the following paragraph and drawing:

    a. The surface, or surfaces, of a feature shall not extend beyond a boundary (envelope) of perfect form at MMC. This boundary is the true geometric form represented by the drawing. No variation is permitted if the feature is produced at its MMC limit of size.

INDIVIDUAL SIZE FEATURE

    b. Where the actual size of a feature has departed from MMC toward LMC, a variation in form is allowed equal to the amount of such departure.

    c. The actual size of an individual feature at any cross section shall be within the specified tolerance of size.

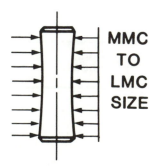

MMC TO LMC SIZE

FIGURE NO. 2–1

In Figure 2-2 is an explanation of Boundary of Perfect Form at MMC. An explanation of how size controls form of the individual feature is also derived according to Rule #1 (under Variations of Form). The pin could taper or bow (hole the same manner) so long as it did not violate the Boundary of Perfect Form and at any cross-section was within the size tolerance limits.

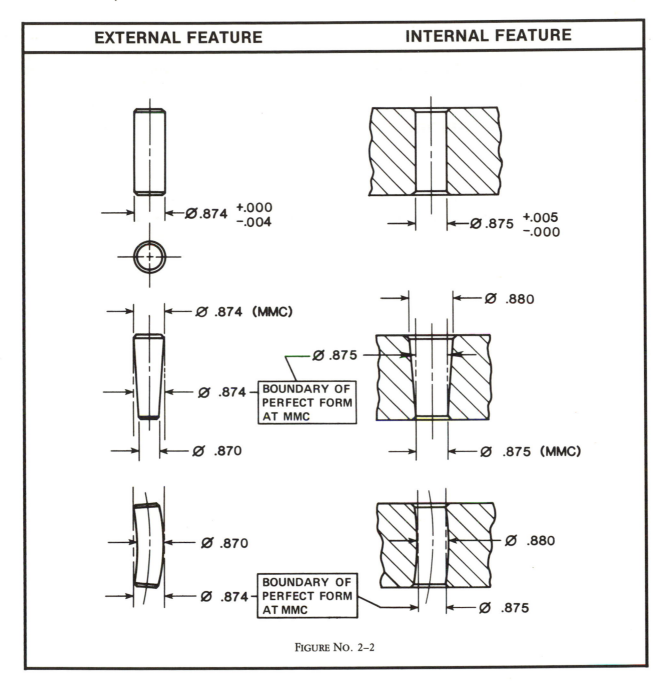

FIGURE NO. 2–2

NOTE: For additional explanation and application see Supplemental Information pages 184-186

MAXIMUM MATERIAL CONDITION (MMC)

The theoretically perfect size (extreme limit, worst case) has a name. It is called the ''Perfect Form at Maximum Material Condition (MMC) Boundary.''

Definition

Maximum Material Condition (MMC) is the condition in which a feature of size contains the maximum amount of material within the stated limits of size (for example: minimum hole diameter, maximum shaft diameter).

LEAST MATERIAL CONDITION (LMC)

Definition

The Least Material Condition (LMC) is the condition in which a feature of size contains the least amount of material within the stated limits of size (for example: maximum hole diameter, minimum shaft diameter).

QUIZ-EXERCISES
MEANING OF A SIZE DIMENSION AND TOLERANCE

The following questions are relative to the material in this chapter. Read the question and answer to the best of your ability. The answers can be found in the companion manual *Answer Book and Instructor's Guide for Introduction to Geometric Dimensioning and Tolerancing*.

Consult your instructor if you have any questions.

1. A size tolerance of an individual feature of size (such as a shaft) controls which two below?

    _____ The cross-sectional elements must be within the size tolerance

    _____ The perfect form at MMC boundary

    _____ The size tolerance and length tolerance relationship

2. What is the MMC size of the shaft below?

Ø.500±.005

.505
MMC Size

3. What is the MMC size of the hole below?

Ø.520±.005

.515
MMC Size

4. On a feature of size, the size controls ___*Form*___ of the feature as well as size?
   (what?)

5. In question 2 above, what is the Perfect Form at MMC Boundary of the shaft shown? _____

6. What is the principle or rule which forms the authority for the answer to question 5?

   _____

7. The limits of size of an individual feature controls perfect form at MMC between features. True or False? _____

8. What is the LMC size of the hole shown in question 3? _____ _____

—————— **NOTES** ——————

3

FORM TOLERANCES

FLATNESS STRAIGHTNESS CIRCULARITY CYLINDRICITY

Form tolerances include flatness, straightness, circularity, and cylindricity.

Form tolerances are applied when features critical to function or interchangeability require specific control, tolerances of size do not provide required control, or other geometrical tolerance controls are to be refined.

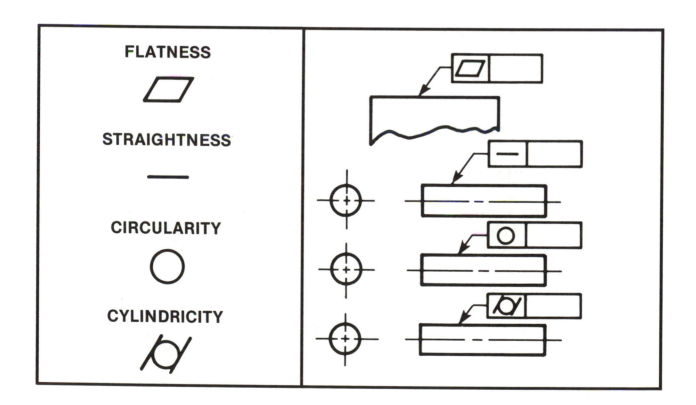

FLATNESS ◻

DEFINITIONS

Flatness is the condition of a surface having all elements in one plane.

Flatness tolerance specifies a tolerance zone defined by two parallel planes within which the surface must lie.

A FLATNESS TOLERANCE

A flatness tolerance is usually based upon a design requirement to provide a precision surface. Where a size tolerance is not precise enough to ensure the surface involved will meet functional requirements, a flatness tolerance can be added as shown in Fig. 3-1.

A flatness tolerance should be less than the total size tolerance involved. Otherwise, there is no need for it as the size tolerance will be adequate. There may be exceptions on parts where ratio of cross-sectional size to its length (long, thin parts) may be excessive or the part may be considered non-rigid. In such cases, special notations or considerations may be necessary.

The flatness tolerance applies to the entire surface unless indicated on a rate basis (flatness per inch, etc.).

FLATNESS TOLERANCE ZONE

A flatness tolerance zone is defined above and shown in Fig. 3-1 under Meaning. Note that a tolerance zone may be imagined as a "window" within which the resulting surface must be placed. Flatness is a requirement which compares the theoretically perfect surface desired (a plane) which cannot be reproduced, to an actually produced (imperfect) surface within a tolerance zone. The tolerance zone being the permissible deviation of the surface from a plane.

FLATNESS IS A 3-D CONTROL

Flatness is also a 3-dimensional control. Since 3-D controls (such as flatness, cylindricity) are more difficult to achieve, notice should be made of the relative costs and difficulties of producing to such requirements.

However, the design requirement and part function should always be the predominant consideration. Consideration of relative costs and difficulty of production should also be made as practical. Designing around specific manufacturing processes is normally discouraged. Define the end product to be produced, not how it is to be produced.

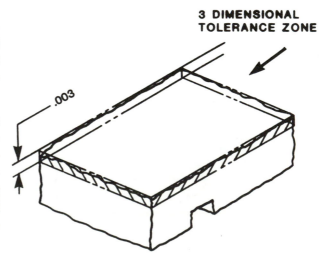

FORM ERROR PERMITTED BY SIZE TOLERANCE

In the introductory chapter, a number of points are made regarding "size" relationship to geometric tolerances. The major theme is that size only controls size and not relationships. We also learned

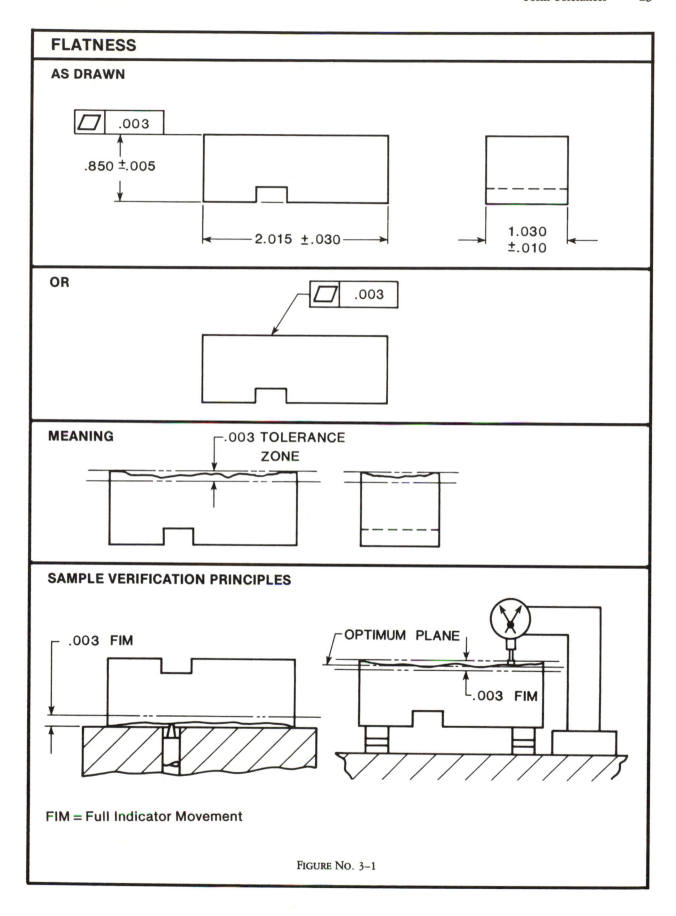

FIGURE NO. 3–1

that size does have a very specific meaning. Rule #1—The Limits of Size Rule—establishes a perfect form at MMC boundary on each individual size feature. However, within that boundary (extreme limit, worst case) as size deviates from MMC size toward LMC size, form deviation to that extent is permitted. For that reason, this effect can also be stated "size controls form of the individual feature" or "form error is permitted to the extent of the size tolerance."

The permitted form error in such instances is, however, a composite of various form errors such as flatness, straightness, circularity, cylindricity. When this composite error (size and form) is permitted, we normally do not care what variety of form error is involved. In other words, if the design does not require anything better in form controls, size is acceptable as is. There is then no need to isolate, in a specific way, any closer precision of the feature.

DEVELOPING A FORM TOLERANCE

From the preceding paragraphs it can be determined that any necessary form control would normally be required to be a refinement (closer than) of the size tolerance.

Explore this idea as follows: Suppose the part below is conceived as a design requirement. The size dimensions are sufficient to control the overall size of the part.

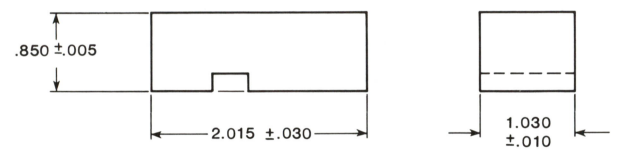

Isolating only the vertical .850 size, a diagram of a part using the size tolerance as a possible form error extreme could appear as:

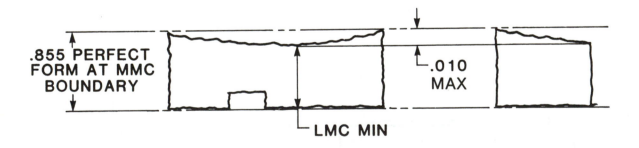

Although the vertical .850 dimensioning is not critical in height, let us assume its form at the upper extremity is. That is, a mating part rests on this surface and greater precision is required than could result through the .850 size dimension tolerance. The designer could then determine that a flatness tolerance on the upper surface is required. A permissible error (tolerance) is decided as .003. It obviously must be less than the already established maximum allowable size tolerance of .010. A logical rule-of-thumb in such situations is to consider the determined flatness should be equal to, or less than, 1/2 of the total size tolerance, such as .003 shown on the following page.

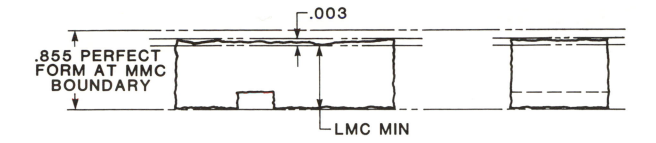

A FLATNESS TOLERANCE HAS BEEN DEVELOPED

The foregoing discussion has explored how a flatness tolerance could be developed. Other methods of determining such a tolerance could obviously be used. This method is representative. The manner in which the flatness tolerance is placed on the drawing and its meaning are shown below and in Fig. 3-1.

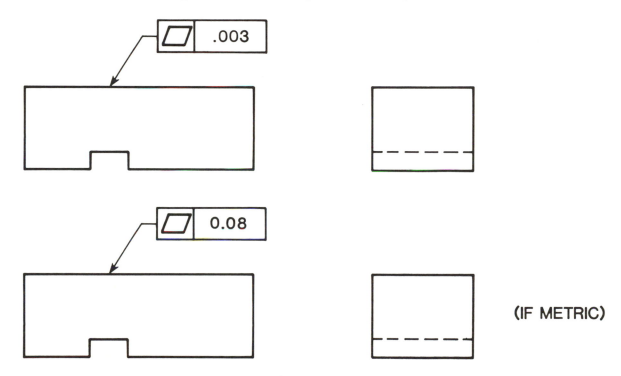

MEANING AND VERIFICATION PRINCIPLES

Since a flatness tolerance is a requirement on the drawing, it must be met. Caution should be exercised, obviously, to be sure that such a requirement is necessary as it must be produced.

The great advantage to production using geometric tolerancing is that it clearly conveys the requirements in an understandable manner. The needed manufacturing processes, tools and equipment can be more readily planned and executed.

Representative verification principles are shown in Fig. 3-1. Special equipment such as a surface plate with mechanical, air, electronic measuring or optimization (leveling-up) open set-up techniques (lower right) are shown. In verification, clarity of the drawing requirements is essential to the inspection process.

STRAIGHTNESS —

DEFINITIONS

Straightness is a condition where an element of a surface or an axis is a straight line.

Straightness tolerance specifies a tolerance zone within which an axis or the considered element must lie.

A STRAIGHTNESS OF SURFACE ELEMENTS TOLERANCE

A straightness of surface elements tolerance is usually based upon a design requirement to provide greater precision of surface line elements than could be expected from the concerned feature's size tolerance. A straightness tolerance is shown in Fig. 3-2. A straightness tolerance should preferably be shown in the view of the drawing where the straight line elements are represented.

FORM ERROR PERMITTED BY SIZE TOLERANCE

In Chapter 2 the limits of control of a size dimension were explained. Rule #1—The Limits of Size Rule establishes a perfect form at MMC boundary on the individual size feature. However, within that boundary (extreme limit, worst case), form deviation is permitted equal to the amount of departure from MMC size toward LMC size. Thus, it can be stated that ''size controls form of the individual feature'' or ''form error is permitted to the extent of the size tolerance.''

The permitted form error in such instances is, however, a combination of possible form errors. The form error might be composed of such specific form errors as flatness, straightness, circularity, or cylindricity. When this composite error (size and form) is permitted as an acceptable size tolerance, we usually do not care what variety of form error is involved. In other words, if no design requirement causes a need for some specific control which is closer than might result with only size control, nothing more is needed. If, however, there is some need for greater precision, such as a form tolerance, it must be specified.

DEVELOPING A FORM TOLERANCE

If a form tolerance is necessary on a part to comply with a design requirement, its value would normally be less than the total size tolerance. Otherwise, there would be a question as to why the form tolerance was added if it did not refine that control already on the drawing (the size dimension).

The reasoning involved in developing an appropriate form tolerance is shown below. Suppose a cylindrical part is required as follows:

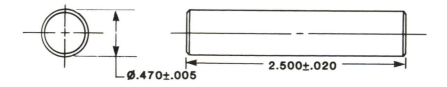

STRAIGHTNESS OF SURFACE ELEMENTS

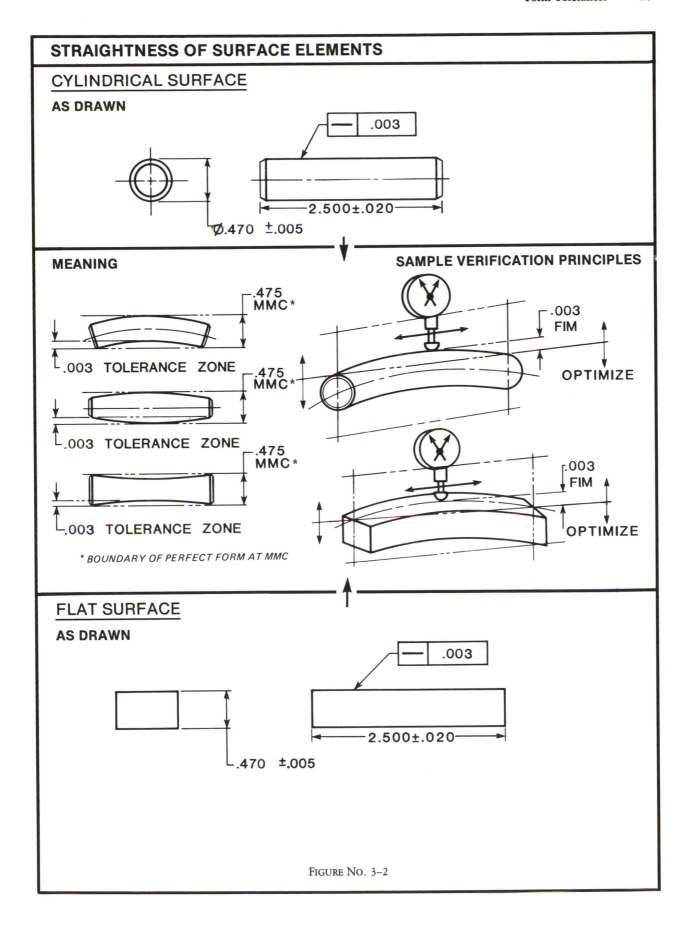

CYLINDRICAL SURFACE

AS DRAWN

⌀.470 ±.005

2.500±.020

— | .003

MEANING

.475 MMC *

.003 TOLERANCE ZONE

.475 MMC *

.003 TOLERANCE ZONE

.475 MMC *

.003 TOLERANCE ZONE

* *BOUNDARY OF PERFECT FORM AT MMC*

SAMPLE VERIFICATION PRINCIPLES

.003 FIM

OPTIMIZE

.003 FIM

OPTIMIZE

FLAT SURFACE

AS DRAWN

— | .003

2.500±.020

.470 ±.005

FIGURE NO. 3–2

Dealing only with the concerned ⌀.470 ± .005 size (the individual size feature), a diagram of the above part using the size tolerance as possible form error extremes, could appear as:

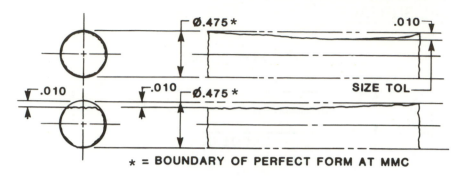

* = BOUNDARY OF PERFECT FORM AT MMC

Note that this could result in "form" errors of the straightness of the lengthwise (longitudinal) line elements of the cylindrical surface up to .010, could permit circularity errors at cross-sections of up to .010 and also could permit out-of-parallelism of opposed elements (deviation from cylindrical shape) towards a tapered or sloped surface of up to .010. All of this could be permitted as acceptable composite variations within the size tolerance, ⌀.475 ± .005. The perfect form at MMC boundary and limits of size tolerance restrict such error within the .010 total size limits of the feature.

In this instance, let us assume that the size tolerance of the ⌀.475 feature can be ± .005 and be suitable to the design requirement. It is not too critical. However, a consistency of the "form" of the part is required wherever it is within the size tolerance. The next question is: Which "form" control is to be isolated as the most important of those referenced above? Is the design requirement primarily concerned with straightness, circularity, or some type of control of opposed elements (restricting taper, etc.)? The designer must determine this as based upon the design details and the part function.

Suppose, in this example, that straightness of the lengthwise (longitudinal) surface elements is determined as the primary concern. We are thinking of all of the elements but must indicate such a control (straightness) as if it were each element one-at-a-time. A straightness tolerance of .003 is determined as the maximum acceptable to the part function. Each such element is imagined as a lengthwise one varying within a tolerance zone (a two dimensional zone). This tolerance must be in a plane (that's why it's called two dimensional) represented as a cutting-plane through the nominal axis of the part. This rather confusing explanation can be cleared up easily by observing the drawing below:

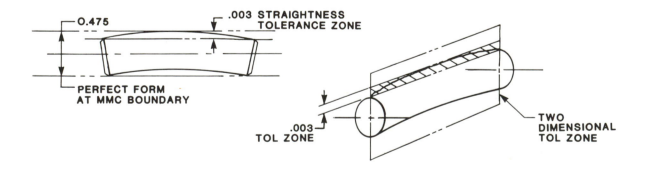

The selected straightness tolerance must be less than the concerned feature's total size tolerance. In this case, the total size tolerance is .010 as noted earlier. Thus, a value less than .010 must be used in such a situation. A good rule-of-thumb which can be used in helping to determine the maximum allowable straightness tolerance is to restrict it to a value which is equal to, or less than, 1/2 that of the concerned size tolerance. In this case, .003 was selected. If this restricted tolerance does not appear necessary to the design, then a straightness tolerance is probably not necessary. Using logic of this kind will prevent over-use of form tolerances and misuse of the system.

If a .003 straightness of surface elements is selected, the drawing below illustrates how the straightness tolerance zone relates to the perfect form boundary (the LMC size limits) and is a refinement of the .010 total size tolerance. The drawing further shows that the straightness tolerance controls only straightness of the surface elements. The part could yet be tapered and out-of-circularity within size limits. However, the chances are good that some limitation of these possible errors will result because of the manufacture of the .003 straightness tolerance.

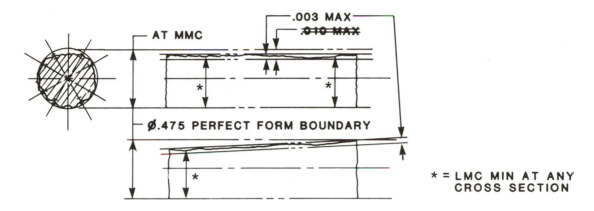

A STRAIGHTNESS OF SURFACE ELEMENTS TOLERANCE HAS BEEN DEVELOPED

The foregoing discussion has explored how a straightness tolerance of surface elements could be developed. The manner in which the straightness tolerance is placed on the drawing is shown below.

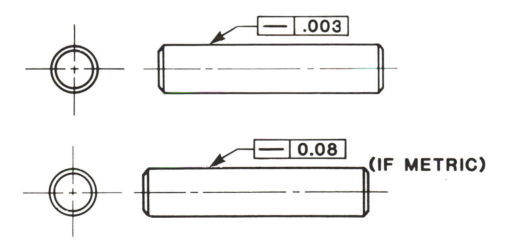

THE DESIGN REQUIREMENT

The preceding paragraphs explain the reasoning in determining the need for a straightness of surface elements tolerance and how it is established and placed on the drawing. As seen in Fig. 3-2, the design requirement has been clearly stated for uniform interpretation. Production of the part would proceed with the appropriate manufacturing processes and tools. The manufacturing operations would be established by a clear understanding of what is to be done. The biggest assist the drawing can be to production is to provide an easily understood meaning.

MEANING AND VERIFICATION PRINCIPLES

Note under Sample Verification Principles in Fig. 3-2 that the surface elements are controlled, whether the part surface error is bowed, barrelled or waisted, up to .003 FIM. Any combination of such errors would be assumed possible. Rule #1—Perfect Form at MMC Boundary restricts the part outer limits and the LMC limit at any cross section of the part must not be exceeded.

The verification process would first require that the part be stabilized (held, mounted) so a measuring device (dial indicator) can establish a zero basis. This is often called "optimizing" (leveling to the measuring table). Then, with the part stationary (not rotating) the measuring device determines, with longitudinal (lengthwise) movement, what error that particular surface line element of the part has. It must be within the .003 FIM. Further similar sample measurements, selected at the discretion of inspection, would be made on other lengthwise elements of the part. All checks made must comply within the .003 straightness tolerance. The part must, of course, also meet size requirements.

OTHER SURFACE ELEMENT STRAIGHTNESS APPLICATIONS

Straightness of surface element control can be applied to any surface which has straight line elements (flat, conical, sloped). Fig. 3-2 at lower right shows straightness applied to a flat surface. Essentially the same reasoning can be applied to such surfaces as has been discussed on the line elements of the cylindrical surface. Under Sample Verification Principles, Fig. 3-2, a measuring process similar to that used previously is described. Also, similar meaning is found on the flat surface part as shown in the middle left of the drawing. The knowledge and skills of the inspector are required to select appropriate elements within the intent of the callout.

STRAIGHTNESS OF AN AXIS TOLERANCE —

The previous section on straightness tolerance addressed the situation where longitudinal (lengthwise) surface element error was the primary concern. That is, the composite error of the surface elements, as possible deviations within the stated size tolerance, were not acceptable to the design requirement. Therefore, a surface element straightness requirement was added as a refinement to fulfill the design intent.

Depending upon the part function and its diameter to length ratio, straightness of another variety (straightness of an axis) may be desired.

FORM ERROR PERMITTED BY SIZE TOLERANCE

As discussed earlier, Rule #1—The Perfect Form at MMC Boundary restricts any form error to that which could possibly occur within the limits of size. This applies to those form errors, such as any uncontrolled straightness surface error which results from size deviation or a straightness of surface element tolerance which is specified.

DEVELOPING A STRAIGHTNESS OF AXIS TOLERANCE

Where certain design requirements call for a longer part, that is, where the ratio of the diameter of a cylindrical part to its length may be excessive, such a part may need to be controlled by a straightness of an axis tolerance. A cylindrical part which is 10 times or more its diameter in length, using a rule-of-thumb to identify such a part feature, could possibly require a straightness of axis control.

PERFECT FORM AT MMC BOUNDARY EXCEEDED

Where such a part as previously described is to be considered, the excessive ratio, length to diameter, could cause a situation in which Rule #1—Perfect Form at MMC Boundary is impractical. That is, the part in its production may not find it possible to comply with the restrictions of perfect form at MMC. Since such a part is longer, its ability to stay within the same limits of the perfect form boundary as a much shorter part is often not possible. The situation is, therefore, different and may require special consideration.

Consider the part below. The reasoning involved in developing an appropriate straightness tolerance is shown. Suppose a cylindrical part is required as follows:

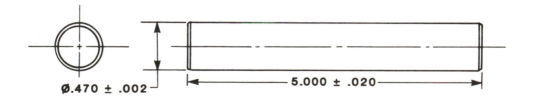

Ø.470 ± .002   5.000 ± .020

Suppose that the diametrical size of the part (Ø.470 ± .002) is critical to its strength and function. Due to excessive length to diameter ratio of the shaft, however, the perfect form at MMC boundary may not be an achievable limit along its length as shown on the following page. The boundary could be exceeded as a result of production and the designer must take this possible error into account as it relates to the part function. The kind and amount of precision that is to be required must be decided and specified on the drawing. Otherwise, the results of production on a part of this kind are left to chance and the resulting assembly of the part may not be satisfactory.

From the drawing, the possibility of exceeding the perfect form boundary is seen. The effect on the design requirement must be considered. We can determine the maximum permissible error by setting our own boundary. This selected boundary would probably be decided by the amount of part distortion permissible yet which will avoid interference, out-of-balance, etc., in the part assembly and function. If the part revolves in its function, the permissible error (tolerance) could most logically be assumed at the axis (the axis of rotation, center-of-balance, the common denominator of the part, etc.). If the part is not rotational in its function, a similar deduction could possibly also result in the same design consideration. In that case, the part has a surface of revolution even though it may not rotate and the considerations could end up very much the same.

An established maximum outer boundary is determined, for example, of Ø.487 on this part as based upon the design considerations. The drawing shows this condition and also simultaneously relates the part maximum permissible straightness error at the axis. The amount is deduced from the simple calculation shown. The new boundary of Ø.487 now restricts the part error. Since its MMC size is Ø.472, the resulting maximum straightness is Ø.015 at the axis.

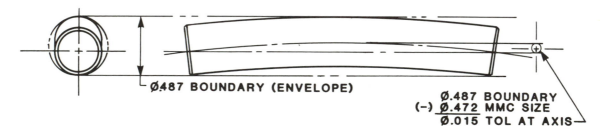

Note, in an example such as this, that the straightness tolerance (of an axis) has exceeded the size tolerance range of .004 maximum. It was the nature of the part which did not respond to the usual guidelines. However, the size limits of the part (Ø.470 ± .002) must be maintained at any cross section as a design requirement.

A STRAIGHTNESS OF AN AXIS TOLERANCE HAS BEEN DEVELOPED

The preceding discussion has explained how a straightness of an axis tolerance could be developed. Of course, other reasoning and criteria for determining such a tolerance are also possible. The manner in which the developed straightness tolerance is placed on the drawing, its meaning and added detail are shown below.

STRAIGHTNESS OF AN AXIS (RFS) —

A straightness of an axis tolerance is usually based upon a design requirement to control the necessary precision of a longer part and is used where its axis (axis of rotation, center-of-balance , etc.) is the predominant design concern. This type of part and its intended design function is usually further affected by the inherent natural problems brought about by the extended diameter-to-length ratio of the part. Its straightness is then more difficult to achieve. A straightness of an axis example is shown in Fig. 3-3. As shown, the straightness should preferably be shown in the view of the drawing where the axis is seen via the "center line" which is the drawing convention that represents the axis. A straightness of an axis tolerance is usually greater than the size tolerance (although it does not have to be). Due to the nature of such parts, they are treated as an exception to the normal size/form tolerance relationship where size is refined (a smaller value) by form control.

THE DESIGN REQUIREMENT

The preceding paragraphs explain the reasoning for "straightness of axis" and considerations in determining the need, uniqueness, and method of stating the requirement on the drawing. In the feature control frame detail of the straightness of an axis callout in Fig. 3-3, the control is placed with (or closely adjacent to) the size tolerance, $\varnothing.470 \pm .002$. Note also that the diameter symbol ($\varnothing$) has been used in the feature control frame to indicate the tolerance zone is at the axis and is diametrical. This symbol is used throughout the geometric tolerancing system to indicate the tolerance zone is diametrical and at the axis of the concerned feature. The feature control frame placement and the details of the contained symbols and numbers clearly indicate the requirement. The use of this callout automatically removes the Rule #1—Perfect Form at MMC Boundary restriction. The new boundary has a name. It is known as "Virtual Condition."

Virtual Condition is the boundary generated by the collective effects of the specified MMC limit of size of a feature and any applicable geometric tolerances.

Each circular cross-sectional element of the surface must, however, remain within the specified limits of size. The straightness tolerance applies to the entire length of the part unless otherwise indicated.

REGARDLESS OF FEATURE SIZE (RFS)

Regardless of Feature Size (RFS) is the term used to indicate that a geometric tolerance of datum reference applies at any increment of size of the feature within its size tolerance.

Regardless of feature size applies to the part in Fig. 3-3. A straightness of an axis requirement is applied only to size features. It is a unique relationship of the axis of that size feature compared to a true axis. The "unless otherwise specified" meaning of the straightness of an axis callout is implied to be on a regardless of feature size (RFS) basis. This is automatic because there is a rule of general agreement in the authoritative standard, Rule #3, which invokes this principle.

In the development of the straightness tolerance as shown in Fig. 3-3, the $\varnothing.015$ tolerance permitted at the part axis is independent of the size (diameter) and any variation in size. The design requirement developed a maximum permissible straightness of the axis tolerance for the shaft at its worst case relationship in the design requirements. But, no matter to which size (cross-sectional elements) the part is produced, the $\varnothing.015$ is the maximum limit of straightness tolerance. The effect of size deviation can have no effect on the straightness tolerance as a design requirement. In standard geometric tolerance language, regardless of feature size, the straightness of the axis tolerance is $\varnothing.015$. RFS is automatically invoked by Rule #3.

RULE #3—OTHER THAN POSITION TOLERANCE RULE

RFS applies, with respect to the individual tolerance, datum reference, or both, where no modifying symbol is specified. MMC must be specified on the drawing where it is required.

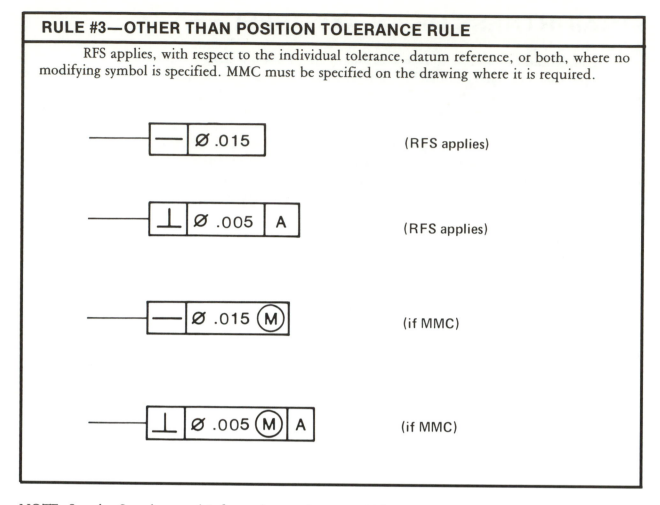

(RFS applies)

(RFS applies)

(if MMC)

(if MMC)

NOTE: See also Supplemental Information section page 186.

MEANING AND VERIFICATION PRINCIPLES

A straightness of an axis is not a simple requirement to meet. However, since there is a clear statement of the requirement, production can proceed with greater confidence via improved communication. The designer has clearly recognized the uniqueness of this type of part and has addressed it for uniform understanding and results.

Note under Sample Verification Principles that the part must be stabilized in an inspection set-up with the lengthwise surface elements of one side optimized (leveled up) to a measuring surface or reference. After sampling one side (the lengthwise element), the part is stabilized again in the same manner on the opposite side (approx. 180 degrees opposite). Readings from the measuring process at the opposed lengthwise surfaces will drive the error at the axis. All results must be within the ∅.015, RFS. If the part is determined as bowed predominantly, direct measurements (each one at a lengthwise element of the surface) can be used. If the part is barrelled (thick in the middle) or waisted (thin in the middle) or a combination of bow and the latter, differential (the difference of the readings of one side versus the other) lengthwise measurements are used. Size at any cross-section (circular element) must meet the ∅.470 ± .002 requirement.

STRAIGHTNESS OF AN AXIS (RFS)

AS DRAWN

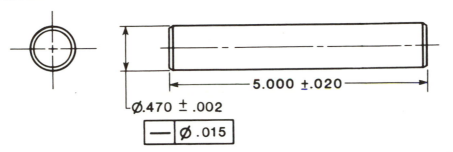

5.000 ±.020

⌀.470 ± .002

| — | ⌀ .015 |

MEANING

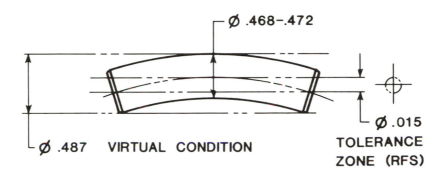

⌀ .468–.472

⌀ .487   VIRTUAL CONDITION

⌀ .015
TOLERANCE
ZONE (RFS)

SAMPLE VERIFICATION PRINCIPLES

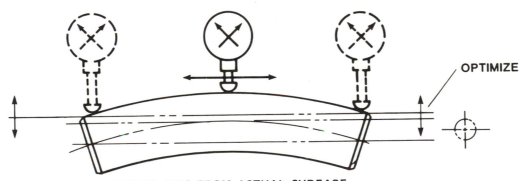

OPTIMIZE

DERIVE AXIS FROM ACTUAL SURFACE
LONGITUDINAL ELEMENTS (RFS)

DETERMINE RESULTANT SHAFT AXIS ERROR WITH DIRECT MEASUREMENTS (IF PART
IS BOWED) OR DIFFERENTIAL MEASUREMENTS (IF PART IS WAISTED OR BARRELED);
MUST BE WITHIN ⌀.015, RFS.

FIGURE NO. 3–3

STRAIGHTNESS OF AN AXIS (MMC) —

Preceding coverage of straightness of surface elements and straightness of an axis (RFS) addressed the subject as it can be applied to achieve required precision of pins, shafts, bores, etc. Some of the reasoning and methods of specification in such applications have been explored.

THE DESIGN REQUIREMENT

Suppose that a part similar to those discussed earlier in this section, such as a shaft or pin, needs somewhat comparable precision but also requires a controlled fit with a mating part hole along its entire length. As shown below at left, a maximum size shaft of $\varnothing.472$ would fit into a minimum size hole of $\varnothing.482$. However, as was emphasized earlier in the diameter-to-length ratio situation, a similar situation can exist in this situation. A 5 to 6 times diameter-to-length ratio can serve as a rough rule-of-thumb to identify such a part feature where an out-of-straightness should be added to the consideration. If a $\varnothing.470 \pm .002$ size tolerance is used as in the example, the middle example reveals that the shaft could be out-of-straight at the axis $\varnothing.010$ with the part at the high (MMC) limit of size $\varnothing.472$. In that case the .010 clearance is used up in form (straightness) error but is yet an acceptable part. It would fit. Size and form (straightness) together in composite represent fit. In the situation at lower right, however, note that if the shaft deviates from its MMC size (gets smaller at cross-sections) the out-of-straightness could increase that amount (for example to $\varnothing.014$ of shaft LMC size). Yet, the parts still fit under that condition and could be likewise an acceptable relationship between them. The question now becomes: Is this kind of relationship (any of the below) desirable or acceptable to the design requirement?

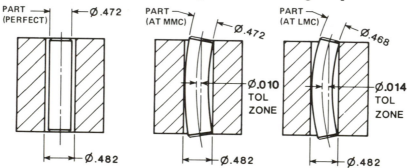

If the above situation is desirable or acceptable to the design requirement, this relationship and dynamics of the shaft straightness can be specified by the MMC principle.

For purposes of this example, the hole in the mating part is assumed as perfect to emphasize the shaft straightness considerations. Of course, perfect parts cannot be made. Therefore, even though the hole with its surrounding material is more stable, considerations could be made as to its straightness requirement if necessary. If done, the worst case of both parts would need to be considered in their relationship.

MAXIMUM MATERIAL CONDITION

The foregoing discussion cites the reasoning and basis for possibly applying the MMC principle on the shaft with geometric tolerance, in this case "straightness-to-an-axis" control. If the design requirement desires, or will permit, the straightness tolerance to increase as the size of the shaft departs from MMC as illustrated above, it can be stated. To specify such a requirement the MMC symbol (M) is added to the feature control frame following the straightness tolerance as shown in the As Drawn example in Fig. 3-4. We have modified Rule #3 to MMC by inserting the (M) symbol. This is the reason this symbol is also called a modifier. See also Supplemental Information, pages 185, 187, and 188.

STRAIGHTNESS OF AN AXIS (MMC)

AS DRAWN

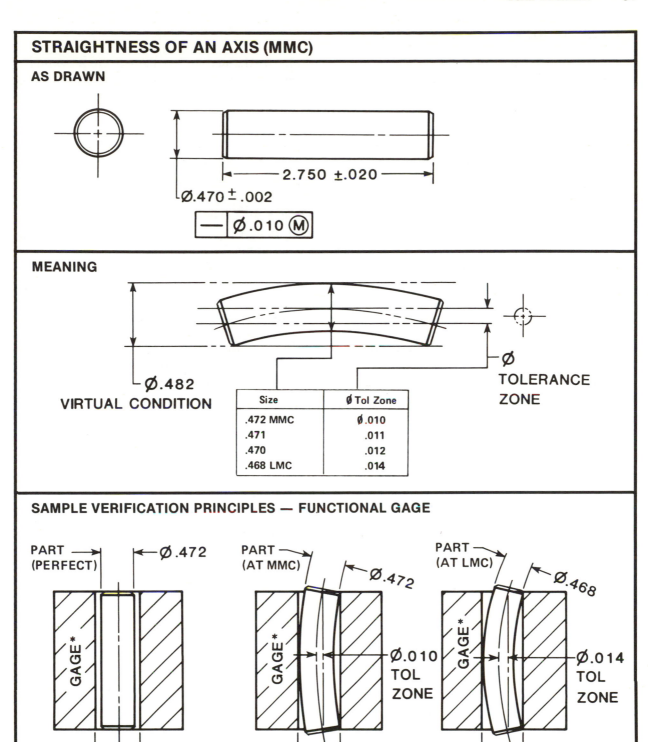

2.750 ±.020

Ø.470 ± .002

— | Ø.010 Ⓜ

MEANING

Ø.482
VIRTUAL CONDITION

Ø
TOLERANCE
ZONE

Size	Ø Tol Zone
.472 MMC	Ø.010
.471	.011
.470	.012
.468 LMC	.014

SAMPLE VERIFICATION PRINCIPLES — FUNCTIONAL GAGE

PART
(PERFECT)

Ø.472

GAGE*

Ø.482♦

PART
(AT MMC)

Ø.472

GAGE*

Ø.010
TOL
ZONE

Ø.482♦

PART
(AT LMC)

Ø.468

GAGE*

Ø.014
TOL
ZONE

Ø.482♦

* GAGE OR REPRESENTATIVE MATING PARTS
♦ VIRTUAL CONDITION SIZE (OF SHAFT) = GAGE HOLE SIZE

FIGURE NO. 3–4

LIMITS OF SIZE

Previously under Rule #1—The Limits of Size Rule, we defined and explained the reasoning of the perfect form at MMC boundary. In that case, MMC was introduced as a way of limiting a size boundary on a worst case size of a feature such as a shaft or pin. However, the MMC principle also can be used as a concept to represent worst case size and form, orientation or location limits. Such application is used only with geometrical tolerances. The example described here is one of those applications. The use of the MMC symbol Ⓜ is required as shown.

MEANING AND VERIFICATION PRINCIPLES

Note under Meaning in Fig. 3-4, a virtual condition results from the collective effect of size and straightness of $\varnothing.482$. This is the exact same value the design was based upon in establishing the straightness tolerance and represents the part worst case at assembly. The tabulation below represents the straightness tolerance zone increase in the amount of the departure from MMC size of the shaft. This, of course, gives production more tolerance to work with and should improve the economics of manufacture. A clear understanding of the requirement also results with evident advantages made possible by MMC Ⓜ. This potential added tolerance is sometimes called a bonus tolerance for obvious reasons.

Under Verification Principles a functional gage is shown. A functional gage is a measurement tool representing the mating part at worst condition. The gage is made to very small gage-making tolerances taken from the part (shaft) geometric tolerance value. Note that the lower figure depicts a gage as a representation of the mating part. The reasoning of the previous discussions is clearly illustrated by the gaging example. Note that the virtual condition size of the shaft establishes the gage hole size nominally. Gage making tolerances are not shown.

If the part (shaft) will drop through the gage hole, the part has met the straightness requirement. Size of the shaft at cross-sections must also meet its stated size limits in a separate check using other measuring tools. If a functional gage is not used, RFS techniques of the previous example would be required.

QUIZ-EXERCISES
FLATNESS AND STRAIGHTNESS TOLERANCES

The following questions are relative to the material in this chapter. Read the question and answer to the best of your ability. The answers can be found in the companion manual *Answer Book and Instructor's Guide for Introduction to Geometric Dimensioning and Tolerancing*.

Consult your instructor if you have any questions.

1. Which of the following characteristics (types of control) would be used to control line elements on a cylindrical or flat surface?

_____ Flatness

_____ Straightness

2. Which of the controls in question 1 is a two dimensional control? _____

Which is a three dimensional control? _____

3. On the drawing below, show how you would specify the accuracy of the lower surface (lower extremity of 1.610 dimension) to allow a total (maximum) tolerance for bow and other surface inaccuracies of .002.

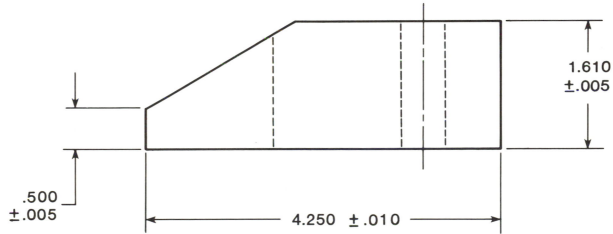

4. Suppose the lower part surface of the figure in question 3 was produced as shown below. Using the form tolerance control selected in question 3, sketch the tolerance zone applicable.

5. On the drawing below, add the requirement that the longitudinal elements of the cylindrical surface are to be within the size tolerance and the boundary of perfect form at MMC basis applies. Use a size dimension of $\varnothing.600^{+.000}_{-.001}$.

6. In your answer to question 5, straightness of the part is controlled to how much as a maximum?

_____

What is the basis for this answer? _____

7. On the drawing below, assume that the part is to mount into bearings and also that the straightness of the longitudinal elements of the cylindrical surface is critical to the design requirements and must be within .0003 total as a refinement of the size control (refer also to questions 5 and 6). Specify this requirement on the drawing.

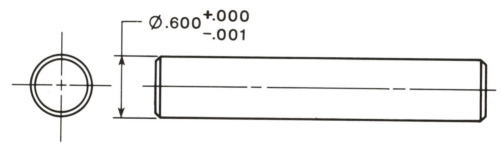

8. What is the maximum permissible straightness tolerance on the preceding part? _____

9. On the drawing below, assume that the part is to mount into bearings but straightness of the longitudinal axis of the cylindrical surface is less critical, to a maximum of $\varnothing.015$ total, RFS, and that Rule #1—Perfect Form at MMC Boundary does not apply. Specify this requirement on the drawing.

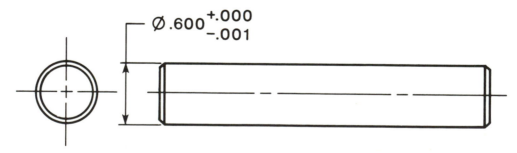

10. The collective effect of the size and form error on the drawing in question 9 results in a virtual condition of _____.

11. Since the .015 straightness tolerance of question 9 was specified on an RFS basis, what is the straightness tolerance permissible if the part size is at $\varnothing.600$? _____ If at $\varnothing.599$? _____

12. Assume the pin shown below question 13 is to assemble with the hole shown. Where interchangeability of parts of this type is required, the condition _____ is often desirable.

13. With pin and hole function fit as a basis (drawing below) the pin can be permitted a straightness tolerance of _____ at MMC exceeding the perfect form at MMC envelope. Show this requirement on the pin.

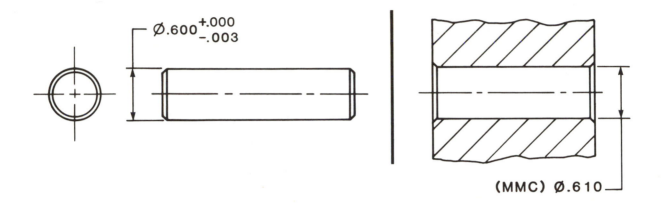

∅.600 +.000 / -.003

(MMC) ∅.610

14. What is the straightness tolerance permissible with the pin of question 13 size at ∅.600? _____ If at ∅.597? _____

15. In the answer to question 14, it is seen that the deviation from MMC size resulted in (added/less) _____ straightness tolerance equal to the amount of departure from MMC size.

16. Straightness tolerance is applicable only to cylindrical parts. True or False? _____

17. A straightness tolerance is normally specified in the drawing view in which the tolerance applies. True or False? _____.

18. Name one type of surface, other than cylindrical, upon which a straightness could be applied. _____.

CIRCULARITY (ROUNDNESS) ◯

DEFINITIONS

Circularity is a condition of a surface of revolution where:

with respect to a cylinder or cone, all points of the surface intersected by any plane perpendicular to a common axis are equidistant from the axis.

with respect to a sphere, all points of the surface intersected by any plane passing through a common center are equidistant from that center.

Circularity tolerance specifies a tolerance zone bounded by two concentric circles within which each circular element of the surface must lie.

Circularity tolerance is a variety of form tolerance. A form tolerance is considered when the size tolerance (only) of the concerned feature does not adequately provide the needed control. Circularity tolerance, when added to the drawing, is a refinement of size controls. Thus, adding a form tolerance requires that it be of more precision (closer tolerance) than the size tolerance. One exception to this is straightness of an axis control where the straightness tolerance usually (not always) exceeds the size tolerance. Circularity tolerance, however, is a refinement of size and, thus, is required to be smaller than the size control.

A CIRCULARITY (ROUNDNESS) OF A CYLINDER TOLERANCE

A circularity tolerance is usually based upon a design requirement to provide greater precision of circular surface elements than could be expected from the concerned feature's size tolerance. A circularity tolerance is shown on a cylindrical shaft in Fig. 3-5. The circularity tolerance callout may be applied in either view of the part, although it is more often placed as shown in the example.

A circularity tolerance may be applied to any part which has circular cross-sections. However, most common application is to cylinders. Conical or spherical features can also be candidates for circularity tolerance. Circularity is a two-dimensional control. Size limits of the shaft and the circularity limits must be met independently.

FORM ERROR PERMITTED BY SIZE TOLERANCE

Form error which is permitted by the size tolerance deviation is a combination (composite) of the effects of such specific form errors as flatness, straightness, circularity and taper. However, they are resultant and uncontrolled errors within the size tolerance and, thus, can be unequally distributed. A circularity tolerance would be added to the drawing when it is necessary to refine and uniformly distribute the permissible tolerance.

CIRCULARITY (ROUNDNESS) OF A CYLINDER

AS DRAWN

⌀ .003

⌀.485 ±.005

MEANING

A

90°

A

A—A

.003 TOL ZONE

⌀.489*

⌀.483*

PART
SURFACE

SECTION A—A

* = SIZE OF TWO CONCENTRIC
CIRCLES WITH PART AT
.489 (AS LARGEST SIZE)

SAMPLE VERIFICATION PRINCIPLES

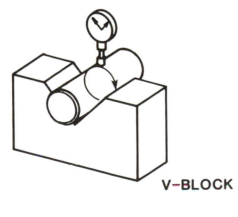

V—BLOCK

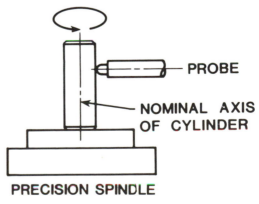

PROBE

NOMINAL AXIS
OF CYLINDER

PRECISION SPINDLE

FIGURE NO. 3–5

DEVELOPING A CIRCULARITY TOLERANCE

If a circularity tolerance is necessary on a part, it would be developed as a refinement of the part size tolerance to comply with a design requirement.

For example, if the shaft shown below did not have a circularity tolerance specified, the figures illustrate the possibilities which could result. The size deviations are form deviations of straightness (lengthwise surface elements), circularity (cross-sectional elements) and taper as shown, possibly to the full extent of the size tolerance .010. The Rule #1—Perfect Form at MMC Boundary restricts the upper size limits of the part and LMC the cross-sectional size limits.

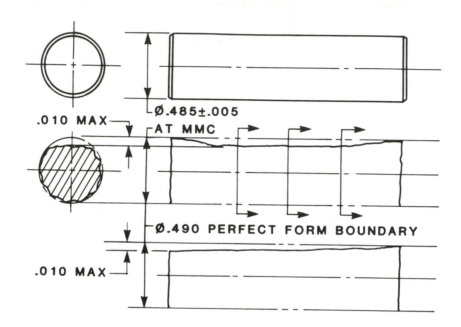

If the predominant form control desired, as based upon design requirements, is to control the circular cross-sectional elements, a circularity tolerance can be specified. Note from the preceding drawing that the uncontrolled circular cross-sectional elements (see arrows) can deviate as an extreme to as much as .010 (see left sectional view). The probabilities are that the extreme error on one side would hardly ever occur. However, there is the possibility that whatever the amount of deviation within size which does occur may not be acceptable to the design requirement. Thus, if the .010 extreme possibility is unacceptable, a circularity tolerance should be determined and specified.

Where a circularity tolerance is specified, it must be less than the stated size tolerance of the concerned feature. Occasional exceptions may occur on non-rigid (flexible) parts. A rule-of-thumb which can be used to determine a circularity tolerance is that it should be equal to, or less than, 1/2 of the size tolerance. Of course, the actual amounts should be based on the design need. That value should be in the range just stated. Otherwise, there can be a question as to whether a circularity requirement is necessary. Size tolerance alone then may suffice.

Suppose that a circularity of .003 is determined as required in the design (see figure below). A refinement of size can then be established and an equal distribution of form tolerance via a circularity tolerance can also then be established. The two concentric circles, as shown below, achieve the equal distribution of form error and replace the unequal distribution possibility which could result from only a size tolerance. See the illustrations below for an explanation of the principles involved.

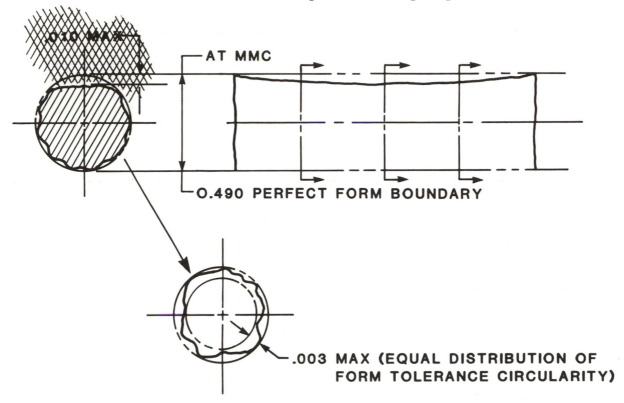

A CIRCULARITY TOLERANCE HAS BEEN DEVELOPED

The preceding discussion has explained how a circularity tolerance could be developed. Other criteria could, of course, be used. The manner in which the developed circularity tolerance may be placed on the drawing is shown below.

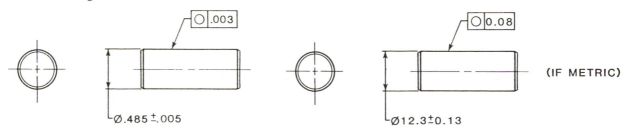

THE DESIGN REQUIREMENT

The preceding pages explain the reasoning in determining the need for a circularity tolerance and how it is established and specified. As is seen Fig. 3-5, the design requirement has been clearly stated for uniform interpretation. The preceding discussion emphasized that the design requirement (what is needed to ensure proper form, fit or function of the part) is the main purpose in adding such a control. Communication from the design to production would proceed with the appropriate manufacturing processes and tools selected.

MEANING AND VERIFICATION PRINCIPLES

The manufacturing operations, as necessary, would be guided by a clear understanding of the part to be produced. Size limits of the part must be met and the circular elements of the produced surface must be within the circularity tolerance zone as well. Circularity is the control of an infinite number (all of them) of physical (actual) circular elements of a circular part such as a cylinder. Each cross-sectional circular element, as produced, must fall within its tolerance zone. Each element is independent of another.

As seen under Meaning in Fig. 3-5, each circular cross-sectional element is imagined as at 90° to the nominal axis of the cylinder. A sample cross-section is illustrated (see A-A). It can be seen that the part surface must meet size limits and also fall within the circularity tolerance zone. As earlier noted, Rule #1— Perfect Form at MMC Boundary remains as an overall size restriction. The LMC limit at any cross section of the part must not be exceeded.

Illustrated under Verification Principles are two types of inspection techniques commonly used in checking circularity. In the V-block method, the part is nested and rotated with a measuring device, such as a dial indicator, in contact with the part. A diametrical value is read which is halved. This result must be within the circularity tolerance. In the case of the example part, .003. Variable angles (the V) found in standard V-blocks (90°, 60°, 120°, etc.) and lobing (raised portions) of the cylindrical part surface place some limitations on high precision measurement with this method. The precision spindle method can ensure precision evaluation of circularity. The part is first mounted on the spindle face. It is then rotated about a nominal (optimum) axis and measurement data is taken via an electronic probe. This data is translated to a polar-graph (circular chart), computer printout, etc. for accurate evaluation.

Other verification techniques are, of course, used.

CYLINDRICITY ⌭

DEFINITIONS

Cylindricity is a condition of a surface of revolution in which all points of the surface are equidistant from a common axis.

Cylindricity tolerance specifies a tolerance zone bounded by two concentric cylinders within which the surface must lie.

A CYLINDRICITY TOLERANCE

Cylindricity, unlike circularity tolerance, can be applied only to a cylindrical feature (i.e. shaft or hole). Cylindricity tolerance, like circularity, is a variety of form tolerance. A form tolerance is considered when the size tolerance (only) of the concerned feature does not adequately provide the needed control. Cylindricity tolerance, as a refinement of size, thus is required to be smaller than size control. A rule-of-thumb which can be used as a guide is: The cylindricity tolerance must be equal to, or less than, $1/2$ of the total size tolerance to be logical.

A cylindricity tolerance is usually based upon a design requirement to provide greater precision of all surface elements together as a single composite control. A cylindricity tolerance is shown on a cylindrical shaft in Fig. 3-6. The cylindricity tolerance callout is usually placed in the view shown in the example although it can be placed in the other (end) view. The size tolerance of the shaft and the cylindricity tolerance are independent requirements.

A cylindricity tolerance is applied only to cylinders (shafts, holes). It is a three-dimensional tolerance and is a composite form control containing straightness of surface elements, circularity of cross sections, and parallelism of opposed elements (taper) as they deviate from a cylinder. All elements must fall within the cylindricity tolerance zone composed of two concentric cylinders.003 apart. Under Meaning, a sample part is shown which is at ∅.824 at the highest elements. This is the outer concentric circle of the .003 tolerance zone. The inner concentric circle is at ∅.818.

FORM ERROR PERMITTED BY SIZE TOLERANCE

Uncontrolled form error which results from size deviation is a combination (composite) of the effects of specific form errors such as flatness, straightness, circularity and taper. These errors which result, however, may be unequally distributed within the size tolerance. Cylindricity is a unique three dimensional control which deals with the form error, straightness of surface elements, circularity and taper all together as one control.

DEVELOPING A CYLINDRICITY TOLERANCE

Suppose that the design requires a short shaft. If, for some reason, there is need to further control form deviation within the size deviation, a cylindricity tolerance may be desirable. Where size can be somewhat lenient, but a consistency of form within that size is necessary, cylindricity could be used. Cylindricity is often required on precision applications where a critical size fit is necessary but further refinement of form is necessary to assure that fit. A bearing journal or bore is a logical possibility for a cylindricity tolerance. Obviously, the precision of size on a bearing application could require a much closer size tolerance than that shown. The values would be of a tighter (closer) variety but the principles would be the same.

CYLINDRICITY

AS DRAWN

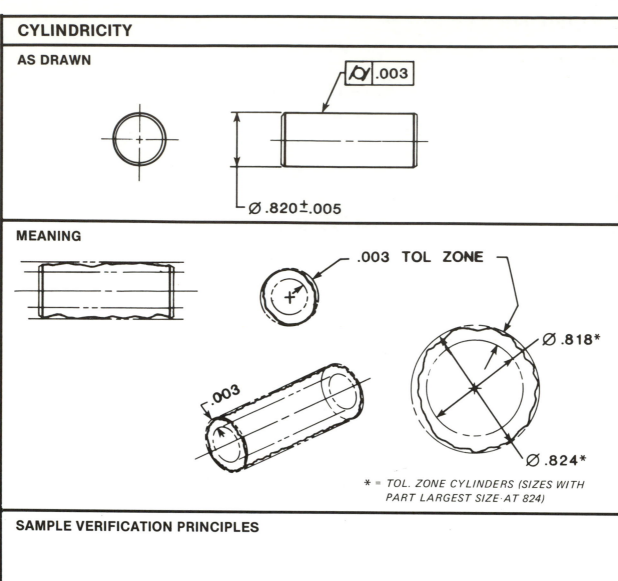

⌀ .820±.005

MEANING

.003 TOL ZONE

⌀ .818*

⌀ .824*

* = TOL. ZONE CYLINDERS (SIZES WITH PART LARGEST SIZE AT 824)

SAMPLE VERIFICATION PRINCIPLES

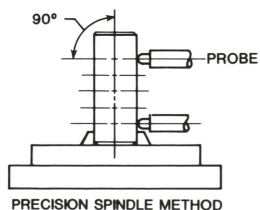

90°

PROBE

PRECISION SPINDLE METHOD

ALTERNATE:
IN OPEN SET-UP, VERIFY CIRCU-LARITY WITHIN .003, STRAIGHT-NESS WITHIN .003, TAPER (CON-ICITY, PARALLELISM OF OPPOSED ELEMENTS) WITHIN .003 PER SIDE.

FIGURE NO. 3–6

If the shaft shown below did not have a form tolerance specified, such as cylindricity, the figures illustrate the possibilities which could result. Size deviations (error) such as straightness (lengthwise surface elements), circularity (cross-sectional elements) and taper could occur up to the full extent of the size tolerance (.010). Rule #1—Perfect Form at MMC Boundary restricts the outer limits of the part and LMC the cross sectional size limits.

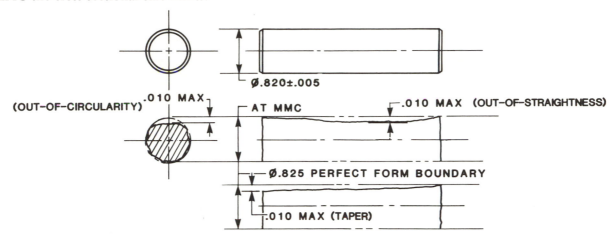

Even though the extreme errors (all to one side) as shown would hardly ever occur, control of the form error as a design requirement could be necessary. That is, some refinement within the size limits may be important to the part function.

Suppose there is a requirement for a refined form tolerance of the shaft within size tolerance as similar to those discussed under straightness of surface elements and circularity in preceding sections. However, in this instance, due to the more critical nature of the part, assume that the part must be controlled relative to all three potential errors — straightness, circularity and taper. The question then is: Should all three controls be placed on the drawing? Answer-probably not, although it would be not be incorrect. Usually, however, when this sort of precision is needed, we are thinking of a combination (composite) control as desirable. If one control can do the job, why use three? It complicates the design and production of the part. The amount of such permissible composite error can be determined and placed directly on the drawing as one requirement with a cylindricity tolerance. It automatically is a composite control of straightness, circularity and permissible taper as deviation from a cylinder.

Suppose that the design requirement on the subject part shaft needs a composite control of all the surface elements such a shown below within a .003 zone as a refinement within size:

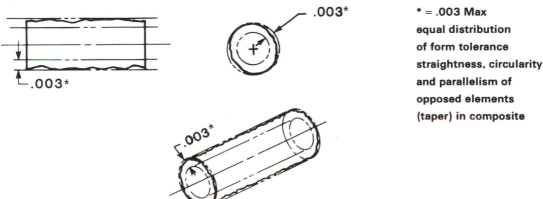

* = .003 Max equal distribution of form tolerance straightness, circularity and parallelism of opposed elements (taper) in composite

Note that establishment of a tolerance zone (.003) which controls all elements, as indicated above in a uniform distribution, ends up as two concentric cylinders .003 apart. From the part shown in the tolerance zone we can derive that it complies with the design requirement. A three-dimensional tolerance zone has been developed in response to the needs. This is a typical cylindricity application.

A CYLINDRICITY TOLERANCE HAS BEEN DEVELOPED

The preceding discussion has explained how a cylindricity tolerance may be developed and justified on more critical or unique design requirements. Caution should be exercised in its selection and use as it is a relatively difficult requirement to verify in inspection. If needed, it is a valid requirement. The manner in which the developed cylindricity tolerance may be placed on the drawing is shown below.

THE DESIGN REQUIREMENT

Preceding pages explain the reasoning and development of a cylindricity tolerance and how it is specified. As is seen in Fig. 3-6, the design requirement has been clearly stated and will be uniformally interpreted. From the preceding discussion, it was noted that the design requirement, to ensure proper form, fit or function of the part, is the main purpose in specifying a cylindricity tolerance. If it is needed, it should be specified. However, as earlier suggested, some caution should be used in selecting cylindricity as it is a relatively difficult requirement to verify in inspection. As is always the case, selection of the proper form controls should be based upon functional requirements and sensible use of the system. When requirements are placed on the drawing, they are requirements to be met. If specified, the cylindricity requirement is clearly communicated to production. Appropriate manufacturing processes and tools would be selected to produce the part.

MEANING AND VERIFICATION PRINCIPLES

Because of its composite nature, a cylindricity requirement is quite readily understood in manufacturing operations. That is, one requirement, rather than three, which might otherwise be suggested (i.e. straightness, circularity, taper) to represent such a control, is better for production. It is the manner in which machining operations proceed normally. A more direct understanding of the end result to be achieved is represented by a cylindricity tolerance rather than bits-and-pieces of a number of requirements to reach an end result.

VERIFICATION PRINCIPLES

Illustrated under Verification Principles (Fig. 3-6) is the representative manner in which a cylindricity tolerance can be verified. Because of its three-dimensional nature, optimization (squaring-up) of the part in the vertical (Z axis) is required in the measuring process. Higher level measuring capabilities and equipment is usually required to adequately verify cylindricity, particularly when high precision parts are being produced. Where such equipment is not available, or not considered necessary, the more standard techniques of checking straightness, circularity and taper individually could be substituted as a last resort. It must be made clear, however, that the latter option does not truly verify cylindricity as the composite tolerance that it is.

3A
QUIZ-EXERCISES
CIRCULARITY AND CYLINDRICITY TOLERANCES

The following questions are relative to the material in this chapter. Read the question and answer to the best of your ability. The answers can be found in the companion manual *Answer Book and Instructor's Guide for Introduction to Geometric Dimensioning and Tolerancing.*

Consult your instructor if you have any questions.

1. Suppose the circular cross sections of a cylindrical part as shown below are critical to a finer degree than the size tolerance would control. What type of form control would be used? _____

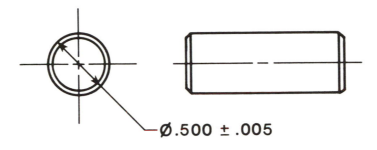

Ø.500 ± .005

2. Show the proper symbolic control on this part using a total tolerance of .002.

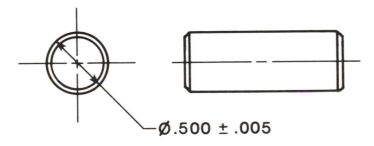

Ø.500 ± .005

3. Show below (sketch) how a tolerance zone would appear if the maximum diameter at that cross-section was Ø.502.

4. Circularity tolerancing can be specified on any part configuration which is _____ in cross-section. Two typical part configurations (other than cylindrical) upon which circularity tolerance may be specified are _____ and _____.

5. Assume that composite surface control of the entire cylindrical surface of the part shown below is required. Add to the figure the proper specification to control the cylindrical surface within .001 total.

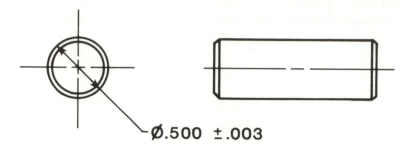

$\emptyset$.500 ±.003

6. Show below (sketch) how tolerance zone is developed on the above part. (Assume the maximum produced size is $\emptyset$.502.)

7. Which tolerance controls are included as a composite form control in cylindricity tolerancing?

_____, _____

_____.

8. Vee block analysis of critical circularity or cylindricity requirements must be wary of which two of the below? (Place a check by your selections.)

_____ size error

_____ diameter versus length

_____ part lobing

_____ variation of V-block angle

9. More accurate analysis methods for verifying circularity or cylindricity can be performed by (place a check by your selection).

_____ gages

_____ precision spindle methods

4

ORIENTATION TOLERANCES

⊥ ∠ ∥

PERPENDICULARITY ANGULARITY PARALLELISM

Orientation tolerances include perpendicularity, angularity, and parallelism.

An orientation tolerance is specified where a relationship of features (surfaces or size features) is required. This relationship may be to achieve a closer orientation tolerance to a size feature, as a refinement of a position tolerance, or as an independent requirement of a size feature. An orientation tolerance is also specified to achieve a relationship between plane surfaces which individual part size tolerances do not control.

Orientation tolerances require datum references.

ORIENTATION TOLERANCES FOR PLANE SURFACES	
For a surface which is related to a datum feature, use one of the following:	
ORIENTATION TOLERANCE	**RELATED PLANE SURFACE**
Perpendicularity (Squareness) ⊥	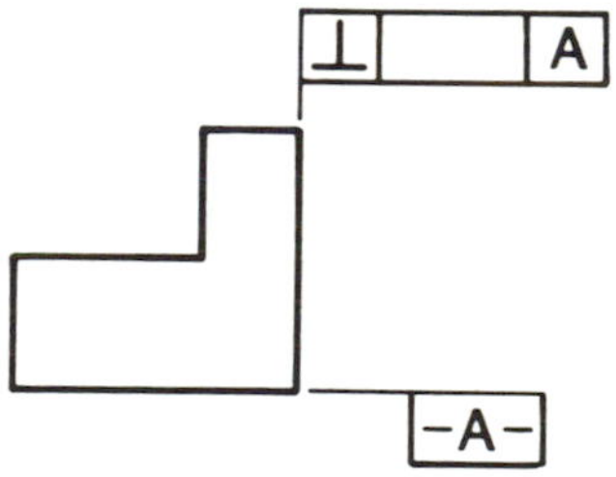
Angularity ∠	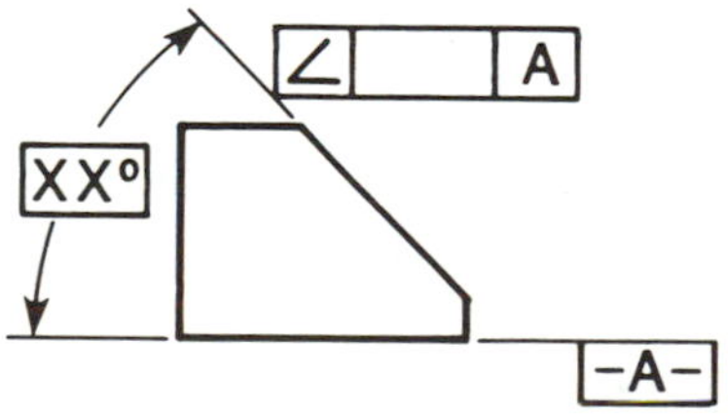
Parallelism ∥	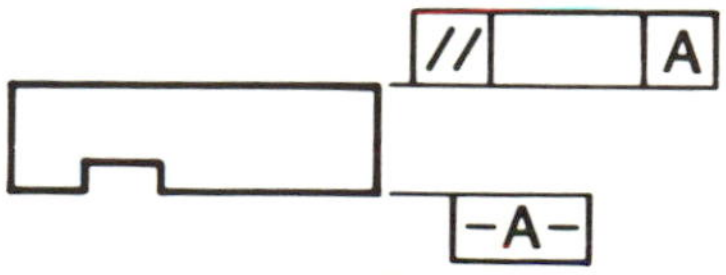

53

Orientation tolerances relate features (surfaces, holes, pins) to one another. Orientation tolerances (perpendicularity, angularity and parallelism) differ from form tolerances (flatness, straightness, circularity, cylindricity) which control the precision of the feature involved only to itself. There is no relationship to other features.

An orientation tolerance can be applied to plane surface (non-size) features or to size features. A distinction between such features was explained under the *Limits of Size* in Chapter 2. It was also noted that a size tolerance of a feature applies only to that individual feature.

RELATIONSHIP BETWEEN INDIVIDUAL FEATURES

The size limits of an individual feature control the form error of that individual feature (Rule #1); but, they do not control the relationship between individual features. Thus, if a relationship between features of some kind requires specific control, it must be determined and specified.

The following quote is found in the authoritative standard:

''The limits of size do not control the orientation, runout or location relationship between individual features. Such features, if shown perpendicular, coaxial, or otherwise geometrically related to each other, must be controlled to avoid incomplete drawing requirements. The use of orientation, runout or location tolerances are then necessary.''

This does not mean, of course, that every relationship shown on a drawing requires a geometric tolerance. Introductory statements on the previous page gives reasons for the need to specify a requirement where related features are involved. Where no orientation tolerance is specified on a part drawing, but a feature is shown perpendicular, angular or parallel to another feature, the following factors will determine the resulting, actually produced, relationship (orientation):

a. **Good workmanship**—the normal skills, tools, and experience of the persons responsible for manufacturing the part will try to achieve an acceptable result. It may be good enough—they will do their best.

b. **Common sense, discretion, good judgement**—the foregoing factor may, or may not, be present and trust is placed upon the manufacturing persons to judge correctly what is necessary to achieve a satisfactory result.

c. **Probabilities**—all of the foregoing items, or none of them, are assisted by probable results. The part may be acceptable under the odds against making bad (non-functional) parts.

d. **Title block or standards**—the ''unless otherwise specified'' controls which may be in force, on the drawing, or applied. The results of manufacturing the parts are controlled as necessary with some guidelines of acceptability. Inspection criteria is established and used if required.

On a typical part drawing, more relationships of feature orientation are implied (shown but not specified) than would be of the specified orientation tolerance variety.

WHEN IS AN ORIENTATION TOLERANCE NECESSARY?

It has been emphasized that a ''size'' tolerance alone does not give adequate control. Where a specific relationship is needed to convey the design requirement, orientation tolerancing is used. It will refine the design details to the necessary degree to ensure that the part has been adequately defined and described. The designer must make the proper judgements. Chapter 1 explained why and where geometric tolerance controls are essential and must be stated.

Defining a part should always start by being sure that the authoritative standard, ANSI Y14.5M-1982, is referenced on the drawing or in supporting documentation. This is essential so that the meaning of size, the rules and all the established symbology, methods, and meaning are uniformly established. Then the proper decisions, selection of geometric characteristics and necessary follow-through can be made with confidence. The standard is our authority and back-up. The standard provides the tools and methods. But we, the user, must provide the judgement as to whether the techniques are to be applied and used; and then, when, which, where, how much, etc.

DEVELOPING AN ORIENTATION TOLERANCE

Where a relationship of features to one another is needed to specify a design requirement, an orientation tolerance may be desirable and necessary. An orientation tolerance is a refinement of the part design which a size tolerance alone will not accomplish. An orientation tolerance may be added to the drawing when a relationship of surfaces to one another or when a refinement of a location tolerance for a size feature (hole, pin) is required. In the latter case, the orientation tolerance is a refinement of the attitude (angle) of the axis or centerplane of the feature within the location tolerance.

DATUM APPLICATION WITH ORIENTATION TOLERANCES

DEFINITIONS

A datum is a theoretically exact point, axis, or plane derived from the true geometric counterpart of a specified datum feature. A datum is the origin from which the location or geometric characteristics of features of a part are established.

Datum, Datum Feature, Simulated Datum

A datum is established from an actual part feature.

A datum feature refers to the actual part feature and, thus, includes all the inaccuracies and irregularities of the produced feature. A datum feature is indicated on the drawing by appropriately attaching or relating the datum feature symbol to the desired feature.

In manufacturing or verification, reference cannot be made from a theoretical plane or axis. Therefore, such a reference is referred to as a simulated datum and is assumed to exist in the precise manufacturing or inspection equipment such as fixtures, gage pins, surface plates, collets, mandrels, etc.

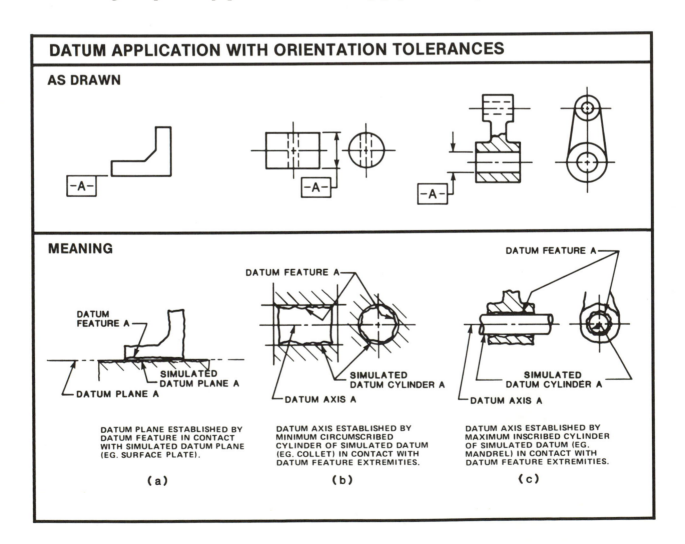

THE DATUM FEATURE SYMBOL

The datum feature symbol specifies the feature(s) of a part from which functional relationships are established. A geometric tolerance is then indicated with respect to that datum feature.

Datum Reference Letter

Each feature requiring identification as a datum on a drawing uses a different reference letter.

Any letters except I, O, or Q may be used (these letters would cause some confusion as they resemble numbers or other symbols). It is common to use the first three letters A, B, C of the alphabet first on a drawing, although it is not required, nor does the choice of letters have any significance. Where the alphabet is exhausted, double letters such as AA, BB, etc. may be used.

The dash lines make the letter stand out in the symbol, and make the symbol distinctive from other drawing conventions. The lines have no other significance.

See also Supplemental Information, page 178.

Placement of the Datum Feature Symbol on the Drawing

The datum identification symbol is placed on the drawing according to standard drawing conventions. Some typical applications are:

AS DRAWN

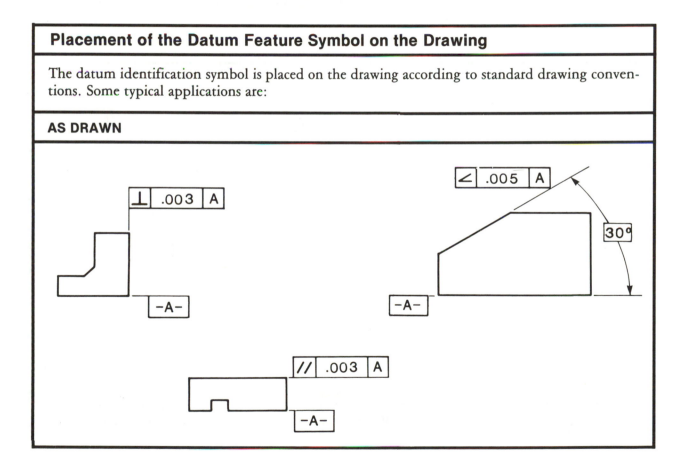

BASIC DIMENSION

A Basic Dimension is used to relate features on a theoretically exact basis. Tolerances for the features involved are established by the associated geometric tolerances.

Definition

A basic dimension is a theoretical value used to describe the exact size, shape, or location of a feature. It is used as the basis from which permissible variations are established by tolerances on other dimensions or notes. A basic dimension is symbolized by boxing it: $\boxed{40°}$ $\boxed{1.270}$

Basic Dimensions Used on Orientation Tolerances

Basic dimensions are necessary on orientation tolerances to relate the concerned features to the appropriate geometry. Since perpendicularity, angularity and parallelism relate features, basic dimensions are involved. However, because of the implied meaning of perpendicularity, with the relationship automatically implied as square, 90° exact (BASIC) and parallelism with the relationship automatically implied basically and exactly parallel (BASIC) no basic dimension need be specified on these controls. The exact desired relationship and stabilization of the tolerance zone relative to the datum is automatically established (as BASIC) by the orientation tolerance callout. Angularity, however, always involves an angle other than 90°. Therefore, it must be specified as a BASIC angle such as $\boxed{40°}$, $\boxed{75°}$ shown on the representative examples below:

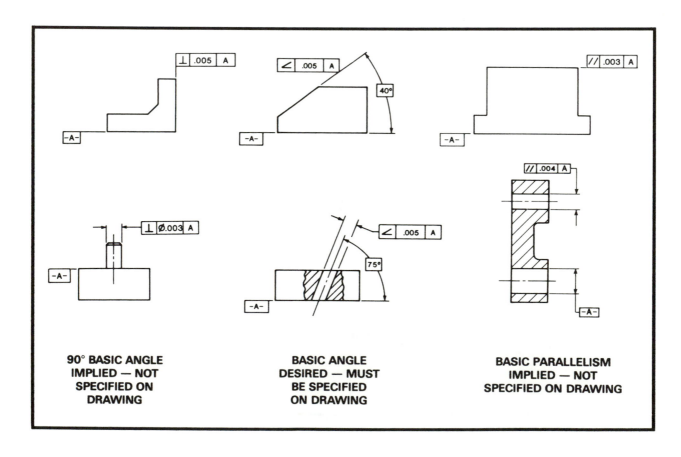

| 90° BASIC ANGLE IMPLIED — NOT SPECIFIED ON DRAWING | BASIC ANGLE DESIRED — MUST BE SPECIFIED ON DRAWING | BASIC PARALLELISM IMPLIED — NOT SPECIFIED ON DRAWING |

PERPENDICULARITY   ⊥

DEFINITIONS

Perpendicularity is the condition of a surface, median plane, or axis at a right angle (90°) to a datum plane or axis.

A perpendicularity tolerance specifies one of the following:

a tolerance zone defined by two parallel planes perpendicular to a datum plane or axis within which a surface or median plane of the considered feature must lie;

a tolerance zone defined by two parallel planes perpendicular to a datum axis within which the axis of the considered feature must lie;

a cylindrical tolerance zone perpendicular to a datum plane within which the axis of the considered feature must lie;

a tolerance zone defined by two parallel lines perpendicular to a datum plane or axis within which an element of the surface must lie.

A PERPENDICULARITY TOLERANCE TO A PLANE SURFACE

A perpendicularity tolerance to a plane surface is necessary where a specific relationship between surfaces is required in the part design. A perpendicularity tolerance to a plane surface example is shown in Fig. 4-1. The perpendicularity tolerance callout should be applied in the view of the drawing where the relationship is best seen. Perpendicularity tolerance is sometimes referred to as squareness tolerance.

A perpendicularity tolerance always requires a datum reference. The datum feature, and thus the datum plane, is required to provide an origin from which the perpendicularity is established. A perpendicularity tolerance applied to a plane surface also controls the flatness of that surface, as well, to the extent of the stated perpendicularity tolerance.

DEVELOPING A PERPENDICULARITY TOLERANCE TO A PLANE SURFACE

Suppose, for example, that the right face surface on the part below (right end of 1.260 ± .010 dimension) is to mate-up with another part or serve some design purpose. This could be where a 90° angle or perpendicularity (squareness) of the surface is critical to fit, function, relationship, appearance, action, etc.

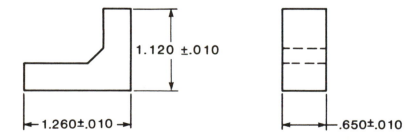

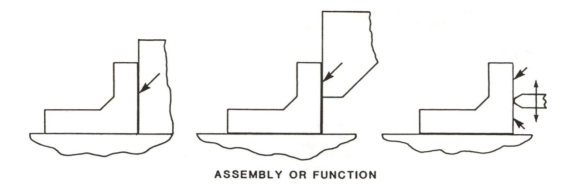

ASSEMBLY OR FUNCTION

The above assemblies suggest some possible purposes the 90° surface may serve in the design requirement.

If, however, we depend upon size control and the factors mentioned before, the precision required for the part may not be achieved. The perpendicularity of the concerned surface may not be adequately controlled. With only size control of the individual features, the example below suggests some of the resulting possibilities (exaggerated to show principles) for the concerned 90° surface.

If the vertical 90 degree surface requires specific control within a specific tolerance zone such as shown below, it can be specified with a perpendicularity tolerance.

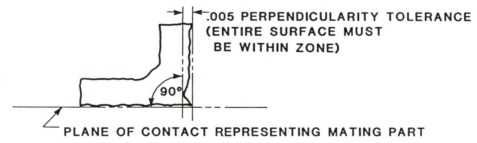

The amount of the perpendicularity tolerance would be determined by the designer and the precision required (for example, a .005 wide zone). This tolerance zone must contain the entire surface. This, therefore, restricts surface error of any variety such as those shown above from angular, to bow, to surface irregularities (flatness). If such a requirement is needed, a perpendicularity tolerance has been determined and can be stated symbolically as below:

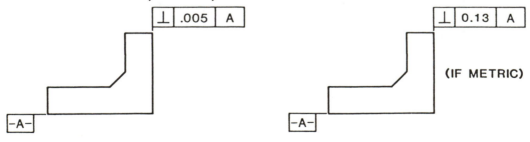

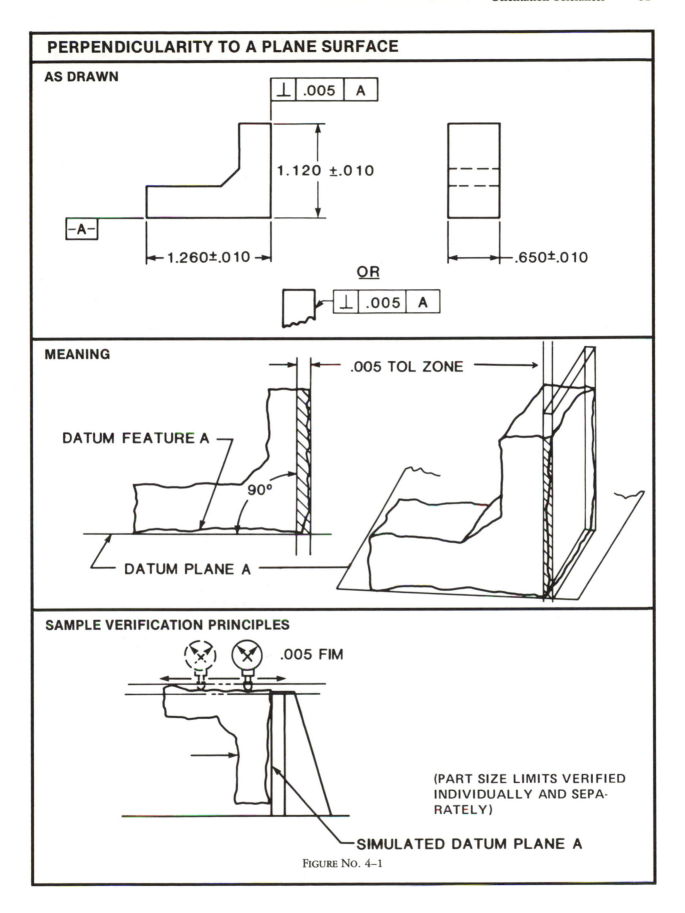

PERPENDICULARITY TO A PLANE SURFACE

AS DRAWN

⊥ | .005 | A

1.120 ±.010

-A-

◄— 1.260±.010 —►

.650±.010

OR

⊥ | .005 | A

MEANING

.005 TOL ZONE

DATUM FEATURE A

90°

DATUM PLANE A

SAMPLE VERIFICATION PRINCIPLES

.005 FIM

(PART SIZE LIMITS VERIFIED INDIVIDUALLY AND SEPA-RATELY)

SIMULATED DATUM PLANE A

FIGURE NO. 4–1

A perpendicularity tolerance to a plane (flat) surface, as discussed and illustrated previously, establishes a tolerance zone (a window) within which the entire produced surface must be contained. That tolerance zone is established as perpendicular (square), at 90° to a datum plane. The datum plane, A, is shown specified on the examples using a Datum Feature Symbol.

The 90° angle is implied as 90° exact, or Basic, as referenced in the geometric tolerancing system. There is no need to specify the 90° Basic angle, however, because it is automatically invoked by the perpendicularity symbolic callout.

DATUM REFERENCE REQUIRED

An orientation tolerance must be related to a datum feature. As shown in Fig. 4-1, the datum A feature (surface) is the feature from which the perpendicularity tolerance zone is related. The surface feature A is a physically produced surface and, therefore, it cannot be perfect (as a plane). However, reference to the datum in the feature control frame actually refers to the datum plane A. The datum A plane, then, is established from the high points (extremities) of the datum feature A as it is brought into contact with the simulated datum plane A. This simulates the mating surface relationship. The simulated datum plane would be the inspection device, such as a surface plate, upon which the part would be rested for verification of the perpendicularity requirement.

THE DESIGN REQUIREMENT

Preceding pages explain the reasoning and considerations in determining a perpendicularity tolerance to a plane surface. Size dimensions of the features involved do not control such relationships. Where a design requirement indicates a need to specifically control a 90° relationship, it can be stated as shown. The value of the perpendicularity tolerance is decided by the precision required in each particular case. The earlier discussions can assist in considering the need for such controls.

MEANING AND VERIFICATION PRINCIPLES

With the requirements clearly stated, the necessary tooling and manufacturing processes can proceed with confidence. The size dimensions and tolerances of the individual features must be met separately as well as the refinement of the feature relationships called for by the perpendicularity requirement. The produced vertical surface must be within the .005 total wide tolerance zone as shown. This applies to the entire surface as a three dimensional tolerance zone.

Illustrated under Verification Principles is a representative manner in which inspection could set-up the part for verification. Mounting the part against a Simulated Datum Plane (in this case an angle-plate) can stabilize the part and permit optimizing (leveling up) of the concerned surface for measurement within the .005 FIM. Other methods of verification, of course, can be used.

———— NOTES ————

_____

_____

_____

_____

_____

_____

ANGULARITY ∠

DEFINITIONS

Angularity is the condition of a surface or axis at a specified angle (other than 90°) from a datum plane or axis. Angularity tolerance specifies a tolerance zone defined by two parallel planes at the specified basic angle from a datum plane, or axis, within which:

the surface of the considered feature must lie;

the axis of the considered feature must lie.

ANGULARITY TOLERANCE TO A PLANE SURFACE

Fig. 4-2 illustrates a variety of orientation tolerance relating features to one another. Angularity tolerance requires the same considerations as perpendicularity tolerance only the desired angle is other than 90°. Therefore, a datum feature reference is required and the desired angle (e.g. 40°) must be stated as a Basic dimension as shown in the example.

Angularity tolerances should be shown in the view of the drawing where the angular relationship to its datum feature reference is clearly seen. An appropriate feature control frame, as shown, is applied to the concerned surface.

DEVELOPING AN ANGULARITY TOLERANCE TO A PLANE SURFACE

Angularity tolerance to a plane surface is a control of a flat surface inclined at an angle other than 90° relative to a datum plane. The permissible deviation is contained with a tolerance zone (a window) stabilized to the basic angle as determined by the design requirement. The example shown in Fig. 4-2 has established that the sloped surface must fall within the .005 wide tolerance zone which is inclined at 40° exact from the mounting face datum A.

The design requirement of angular surface part features and the considerations involved would be similar to those discussed earlier dealing with perpendicularity.

MEANING AND VERIFICATION PRINCIPLES

As is seen in Fig. 4-2, the 40° inclined surface must be produced to fall within the .005 wide tolerance zone. The flatness of the concerned surface is also controlled as part of the angularity requirement. The angularity tolerance must be met as well as all size dimensions and tolerances independent of the angularity tolerance. Most size dimensions on the sample part shown were left off for simplicity and to emphasize only the angularity tolerance and its purpose and meaning.

Appropriate measuring tools or processes would be used by inspection to verify the requirement based upon the principles shown. A clear understanding of the design intent and its meaning will assist in the selected process. Special set-up, fixturing, sine-plate and other comparable methods can be used for verification.

ANGULARITY TO A PLANE SURFACE

AS DRAWN

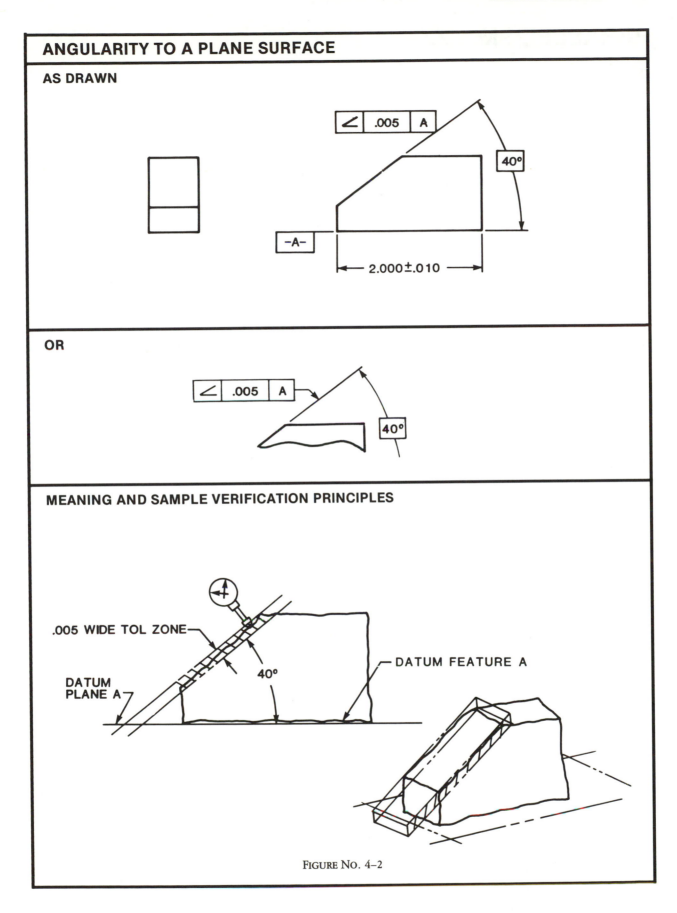

FIGURE NO. 4-2

PARALLELISM //

DEFINITIONS

Parallelism is the condition of a surface equidistant at all points from a datum plane or an axis equidistant along its length to a datum axis.

Parallelism tolerance specifies:

a tolerance zone defined by two parallel planes or lines parallel to a datum plane, or axis, within which the line elements of the surface, or the axis of the considered feature must lie;

a cylindrical zone whose axis is parallel to a datum axis within which the axis of the considered feature must lie.

PARALLELISM TOLERANCE TO A PLANE SURFACE

Parallelism tolerance to a plane surface, as illustrated in Fig. 4-3, is a type of orientation tolerance. It relates features to one another, one being a datum feature. Parallelism tolerance requires similar considerations as discussed in previous examples on perpendicularity and angularity to plane surfaces. However, the relationship is parallelism of one feature to the other (the datum feature). The Basic relationship is parallelism with the tolerance zone stabilized parallel to the datum plane.

The parallelism tolerance may be shown in an appropriate drawing view where the relationship of the concerned features (feature to datum) is clearly seen. A feature control frame is placed on the drawing with one of the accepted methods shown in the desired view.

DEVELOPING A PARALLELISM TOLERANCE TO A PLANE SURFACE

Parallelism to a plane surface is a control of a flat surface and its precision relative to a datum plane. As seen in Fig. 4-3, the permissible deviation is contained within a tolerance zone (a window) stabilized parallel (exactly) to the mounting face, datum feature A and the datum plane A, as established. The example has established that the upper surface must be within the size tolerance ($1.300 \pm .005$) and also, as a refinement of orientation, fall within the .003 wide tolerance zone which is parallel to datum plane A.

The design requirements of parallel surface part features and the considerations involved would be similar to those discussed previously on perpendicularity and the general coverage on orientation tolerances. However, due to the nature of parallelism to plane surfaces, the concerned surface is uniquely involved in both size control and the parallelism control. Therefore, the parallelism tolerance is a direct refinement of the concerned size tolerance and must be less than the size tolerance. A suggested rule-of-thumb in such situations is: The parallelism must be equal to, or less than, 1/2 of the total overall size tolerance involved. Otherwise, there is a question as to the need for a parallelism tolerance. Size control only may be good enough.

MEANING AND VERIFICATION PRINCIPLES

The concerned surface in the example must meet both size and parallelism requirements separately. Note that the flatness of the concerned surface is also controlled as a part of the parallelism requirement.

Appropriate measuring tools and methods would be used in inspecting the requirements as based upon the evident principles shown. Clarity of the design requirement and its meaning will assist in proper verification.

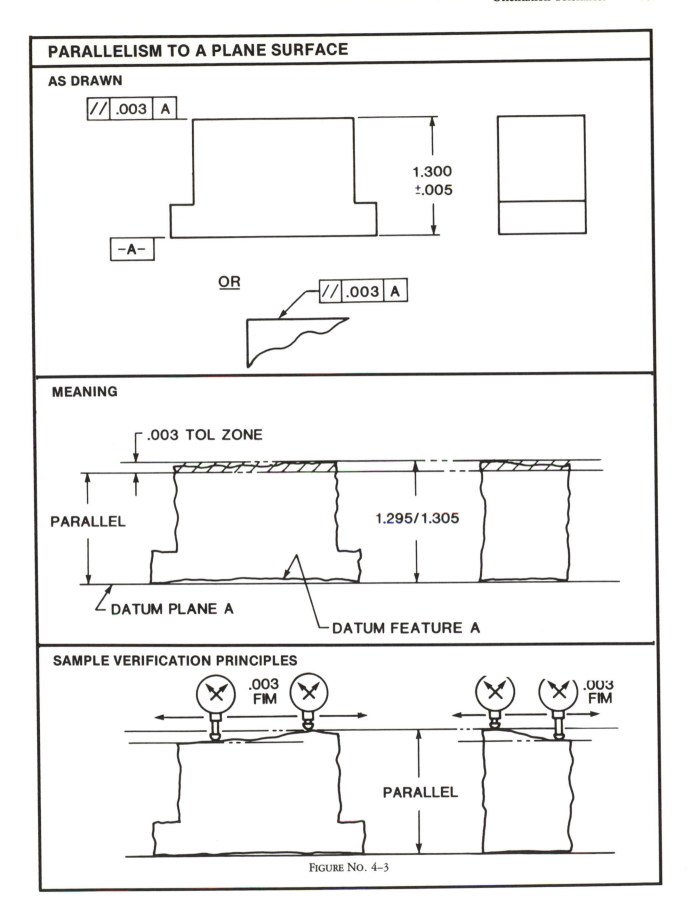

PARALLELISM TO A PLANE SURFACE

AS DRAWN

// | .003 | A

1.300
±.005

–A–

<u>OR</u>

// | .003 | A

MEANING

.003 TOL ZONE

PARALLEL

1.295/1.305

DATUM PLANE A

DATUM FEATURE A

SAMPLE VERIFICATION PRINCIPLES

.003 FIM

.003 FIM

PARALLEL

FIGURE No. 4–3

NOTES

ORIENTATION TOLERANCES FOR SIZE FEATURES

An orientation tolerance can be applied to a size feature such as a hole or pin. The primary purpose of an orientation tolerance used on a size feature would be as an orientation refinement of the locational tolerance of that feature. The orientation tolerance itself is not intended to be used as a location control. It is technically correct to do so, however. If done, (with rare exceptions) it would be a case of applying an orientation tolerance where a positional tolerance should be used. Positional tolerance is designed for the express purpose of locational control. But, it also includes orientation control as part of the location tolerance. If necessary, however, a specific orientation tolerance can be added to improve (refine) the orientation of the feature to its appropriate datum. Obviously such an orientation tolerance would be less (closer tolerance) than the positional tolerance and relative to the same datum reference.

On any of the examples in Fig. 4-4, the hole would require a locational tolerance. A positional tolerance would be the method used if geometric tolerancing techniques were chosen. Only the orientation tolerance feature control frames are shown at left in the examples, however. This is to emphasize orientation tolerance controls at this time and simplify explanation. Later sections will address position tolerance detail.

DEVELOPING AN ORIENTATION TOLERANCE TO A SIZE FEATURE

Suppose, for example, that the hole in the upper part at left is permitted a relatively lenient positional tolerance of Ø.030. It will be noted later, when discussing positional tolerancing, that such a callout would also permit the hole to be out-of-perpendicular to as much as .030, essentially using up the entire (or most) of the positional tolerance. See figure (a) below.

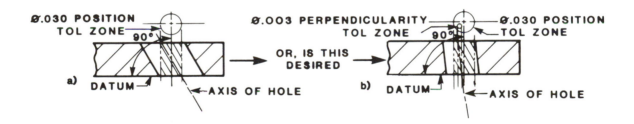

If there is a need to refine the perpendicularity control within the positional control as suggested in figure (b) above, it can be stated by determining and specifying the appropriate datum feature along with the feature control frame on the drawing. For example:

The other examples in Fig. 4-4 represent similar considerations and situations except the orientation relationship could be an angularity or parallelism refinement of a positional tolerance. In all examples, the datum referenced features are necessary as the orientation relationship is from the datum and based on the design requirement.

Orientation tolerance when applied in this manner is usually based upon a situation where the location of the concerned feature is not critical but its orientation is for some functional or assembly purpose. The amount of the orientation tolerance is derived from the design requirement and the size and condition of mating part features, etc.

A DATUM PLANE OR AXIS MUST BE ESTABLISHED

Refer to page 56 for a review of the definition, use and establishment of the datum plane from a plane surface feature. These principles apply to the perpendicularity and angularity examples of Figure 4-4.

Refer also to page 56 for the definitions, use and establishment of the datum axis from a datum feature. These principles apply to the parallelism examples of Figure 4-4. The datum feature is a physical (actual) feature from which the datum axis is established. The parallelism tolerance zone would be parallel to the datum axis, regardless of feature size (RFS). RFS is implied by Rule# 3—The Other Than Position Tolerance Rule.

ORIENTATION TOLERANCES FOR SIZE FEATURES

For a size feature which is related to a datum feature, use one of the following:

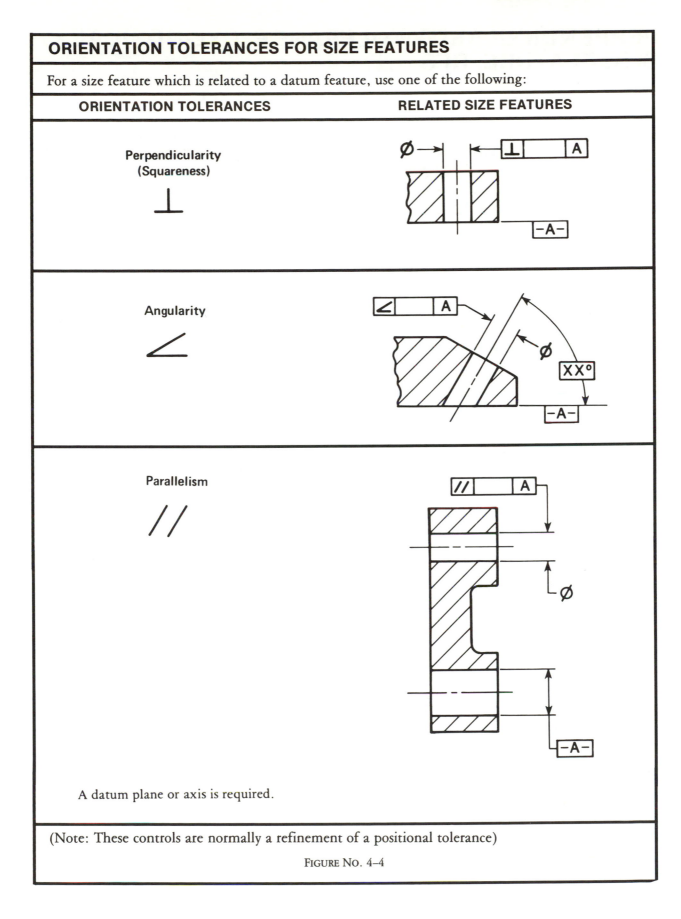

ORIENTATION TOLERANCES	RELATED SIZE FEATURES

Perpendicularity
(Squareness)

Angularity

Parallelism

A datum plane or axis is required.

(Note: These controls are normally a refinement of a positional tolerance)

FIGURE NO. 4–4

PERPENDICULARITY TOLERANCE, CYLINDRICAL SIZE FEATURE, RFS ⊥

As discussed in the preceding section, a size tolerance does not control orientation of that feature. Where a size feature such as a hole or pin is shown on the drawing as square to, or perpendicular to, a surface but is not given an orientation tolerance, it will result as produced in manufacture. The accuracy of the produced feature, insofar as its perpendicularity is concerned, will be guided by the factors listed on page 54 and the locational tolerance specified.

DEVELOPING A PERPENDICULARITY TOLERANCE TO A SIZE FEATURE, RFS

Suppose that the pin shown in Fig. 4-5 is relatively unimportant in its location (position) from the outer surfaces. But wherever it is, it is critical in its perpendicularity to the surface specified as datum A.

The resulting perpendicularity of the pin would be within the positional tolerance if nothing further had been stated on the drawing. If the pin must provide a fit or some precision orientation relationship in the design function, a perpendicularity tolerance could be stated. It would be calculated on the basis of feature relationships and would be a value less than the permitted location tolerance. We would be providing a refinement of the position control relative to its orientation (perpendicularity) to the mounting surface, datum surface A.

A PERPENDICULARITY TOLERANCE TO A SIZE FEATURE PLACED ON THE DRAWING

Where the perpendicularity tolerance calculated (i.e. $\varnothing$.003 as shown) is the maximum amount which can be permitted in the design, it is stated as regardless of feature size. The feature control frame in Fig. 4-5 automatically invokes the $\varnothing$.003 perpendicularity tolerance as RFS by Rule #3—The Other Than Position Tolerance Rule.

RULE #3—THE OTHER THAN POSITION TOLERANCE RULE

For all applicable tolerances other than position tolerance, RFS applies with respect to the individual tolerance, datum reference, or both, where no modifying symbol is specified. MMC (M) must be specified on the drawing where it is required.

The diameter symbol $\varnothing$ is usually included preceding the perpendicularity tolerance value in the feature control frame where a feature of size, such as a pin, is related to a datum surface only as shown on this example. This clearly indicates the tolerance zone is cylindrical.

The feature control frame can be placed in either view whichever is most desirable. If possible, showing the perpendicularity tolerance in the view of the drawing where the relationship is seen can be advantageous to reading the requirement easily.

A datum reference as shown is required in such applications. This stabilizes the tolerance zone functionally to the mounting face, datum A.

See also Supplemental Information, page 188.

PERPENDICULARITY, CYLINDRICAL SIZE FEATURE, RFS

AS DRAWN

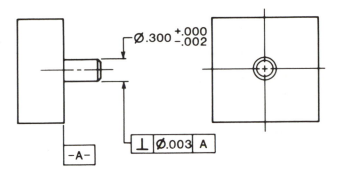

Ø.300 $^{+.000}_{-.002}$

⊥ | Ø.003 | A

-A-

MEANING

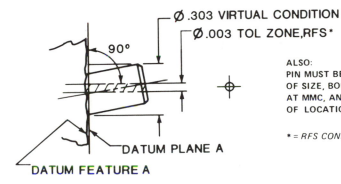

Ø.303 VIRTUAL CONDITION

Ø.003 TOL ZONE, RFS*

90°

DATUM PLANE A

DATUM FEATURE A

ALSO:
PIN MUST BE WITHIN SPECIFIED LIMITS
OF SIZE, BOUNDARY OF PERFECT FORM
AT MMC, AND SPECIFIED TOLERANCE
OF LOCATION

* = RFS CONDITION IMPLIED BY RULE #3.

SAMPLE VERIFICATION PRINCIPLES

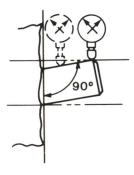

90°

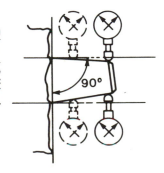

90°

DETERMINE RESULTANT PIN
AXIS ATTITUDE WITH DIRECT
MEASUREMENTS (LEANING
PIN) OR DIFFERENTIAL MEA-
SUREMENTS (CONICAL PIN)
DEPENDENT UPON PIN BASIC
FORM ERROR. RESULTANT
AXIS MUST BE WITHIN Ø.003,
RFS.

FIGURE NO. 4–5

Using the foregoing reasoning, the requirement is placed on the drawing in Fig. 4-5. A representative method is shown. Other alternatives are possible for placing the feature control and datum feature symbols on the drawing as shown below and elsewhere in this text.

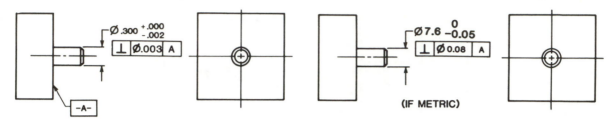

MEANING AND VERIFICATION PRINCIPLES

Under Meaning, the perpendicularity tolerance zone and its Virtual Condition are illustrated. The axis of the pin, as produced, must be within the ⌀.003 tolerance zone, regardless of the pin size (RFS). The pin must also be within the limits of size, boundary of perfect form (Rule #1) and locational tolerance.

Definition

Virtual Condition—The boundary generated by the collective effects of the specified MMC limit of size of a feature and any applicable geometric tolerances.

Under Verification Principles (Fig. 4-5), the inspector must stabilize (hold, mount) the part to its datum plane. Then measurement of the surface elements (sampling them) will determine the resultant axis. If the pin is predominantly leaning (lower left), direct measurements can be taken at the surface to determine whether the resultant axis is within the ⌀.003 tolerance zone. If the pin is other than leaning, such as conical in shape, differential measurements of surfaces elements approximately opposite (180°) are taken. The difference between the opposed readings must be within the ⌀.003 tolerance zone.

PERPENDICULARITY TOLERANCE, CYLINDRICAL SIZE FEATURE, MMC ⊥

As discussed in the preceding section, a size tolerance does not control orientation of that feature. Where a size feature such as a hole or pin is shown on the drawing as square to, or perpendicular to, a surface but is not given an orientation tolerance, it will result as produced in manufacture. The accuracy of the produced feature, insofar as its perpendicularity is concerned, will be guided by the factors listed on page 54 and the locational tolerance specified.

DEVELOPING A PERPENDICULARITY TOLERANCE TO A SIZE FEATURE, MMC

Suppose that an assembly of two parts via a pin and a hole is required. That is, the pin on the lower part (Fig. 4-6) must enter a hole on the upper part with some precision. There is a small amount of clearance between the two features to cause a sort of three-dimensional fit as well as to locate one part to the other. It is critical that the face surfaces bottom-out on each other as the stabilizing surfaces and do not stress (force) the pin.

Observe in Fig. 4-6 under Sample Verification Principles, that if the gage reference was removed and we imagine the gage as a part, the condition described above is illustrated. Assume that this represents the conditions of the mating parts. Note that with both features (hole and pin) perfect in perpendicularity and at MMC (or worst case) with hole at ∅.303 and pin at ∅.300 Fig. a, there would be .003 clearance. This is the designer's determined amount for one extreme case which is acceptable to the assembly and function of the parts. Since perfect parts are not possible (that is perfect in size or perfectly square), the designer also has developed the permissible perpendicularity of the pin relative to the worst case hole (∅.303, Fig. b).

To simplify the example, we have disregarded any concern for the hole's perpendicularity because of its probable better stability. If necessary, we would need to establish its permissible error (worse case) and state the appropriate perpendicularity tolerance as well.

If the designer decides that the reduction in pin size, and thus an increase in clearance or pin tilt is permissible (Fig. c) with no ill effect on the assembly, the MMC principle can be used. Therefore, the amount of clearance between the hole and pin worst case develops the permitted perpendicularity tolerance. By stating the requirement as applying at MMC, the conditions and design function described are captured. That is, as the pin size reduces in size in production (departs from MMC), an increase in the perpendicularity tolerance is permitted.

A PERPENDICULARITY TOLERANCE TO A SIZE FEATURE PLACED ON THE DRAWING

The example at upper right (Fig. 4-6) illustrates an application as described in the foregoing paragraphs. Note that the MMC symbol Ⓜ is stated in the feature control frame and modifies Rule #3— The Other Than Position Tolerance Rule. Note also that the location of the hole, as well as size limits, must be met. Perpendicularity of this kind (to a pin or hole size feature) is a refinement of orientation of the feature within its locational tolerance.

A datum reference as shown is required in such applications. This stabilizes the tolerance zone.

A perpendicularity tolerance of this kind can be shown in either view of the drawing. Preferably, it is shown in the view where the perpendicularity relationship is seen (one can see it better). The diameter symbol ∅ is usually included preceding the perpendicularity tolerance value in the feature control frame where a feature of size, such as a pin, is related to a datum surface only as shown on this example.

MEANING AND VERIFICATION PRINCIPLES

Under Meaning (Fig. 4-6), the perpendicularity tolerance zone is illustrated. The axis of the pin, as produced, must be within the tolerance zone. The tabulation summarizes the resulting available tolerances with given produced sizes as it departs from MMC size. Note that the Virtual Condition $\varnothing.303$ results which represents the worst case relationship of the pin to its mating feature.

One of the advantages to production and verification is the permitted use of a functional gage if desirable. An example is shown in Fig. 4-6. A functional gage can be described as a master mating part. Earlier discussion in this section alluded to this when using this same illustration and figures a, b, c to describe the mating part function. Use of a functional gage can provide economic advantages to production and verification as it reduces inspection time and does not require an inspector's special skills or equipment. It (the gage) also recognizes and permits the extra tolerance sometimes called the bonus tolerance. Note that the gage member (the gage hole) is built to Virtual Condition size. Tolerances for the gage design and construction (not shown) would be determined by the gage designer.

PERPENDICULARITY, CYLINDRICAL SIZE FEATURE, MMC

AS DRAWN

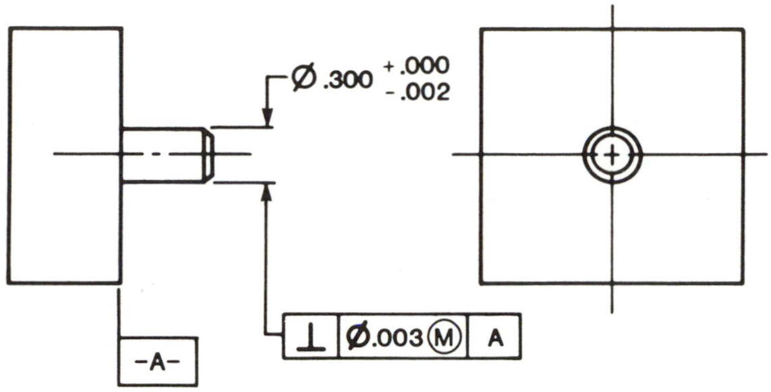

$\varnothing$ .300 $^{+.000}_{-.002}$

⊥ | $\varnothing$.003 Ⓜ | A

–A–

MEANING

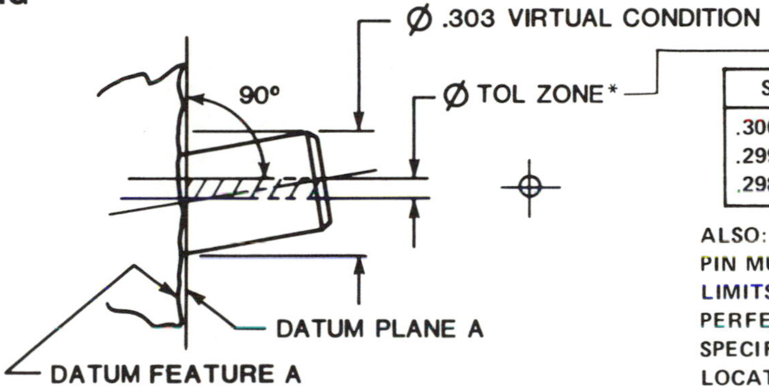

$\varnothing$ .303 VIRTUAL CONDITION

$\varnothing$ TOL ZONE*

90°

SIZE	$\varnothing$ TOL ZONE*
.300 MMC	$\varnothing$.003
.299	.004
.298 LMC	.005

ALSO:
PIN MUST BE WITHIN SPECIFIED
LIMITS OF SIZE, BOUNDARY OF
PERFECT FORM AT MMC, AND
SPECIFIED TOLERANCE OF
LOCATION.

DATUM PLANE A

DATUM FEATURE A

SAMPLE VERIFICATION PRINCIPLES — FUNCTIONAL GAGE

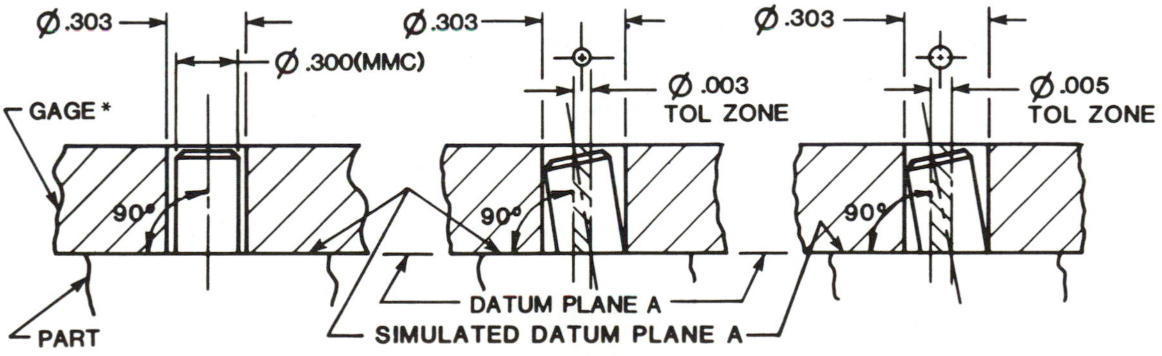

$\varnothing$ .303

$\varnothing$ .300(MMC)

GAGE *

90°

PART

PIN AT $\varnothing$.300 (MMC)
PERFECT PERPENDICULARITY

(a)

$\varnothing$ .303

$\varnothing$ .003
TOL ZONE

90°

DATUM PLANE A
SIMULATED DATUM PLANE A

PIN AT $\varnothing$.300 (MMC)
OUT-OF-PERPENDICULARITY
$\varnothing$.003

(b)

$\varnothing$ .303

$\varnothing$ .005
TOL ZONE

90°

PIN AT $\varnothing$.298 (LMC)
PERMISSIBLE PERPENDICULARITY
$\varnothing$.005 (.003 + .002 SIZE TOL)

(c)

* GAGE OR REPRESENTATIVE MATING PART

FIGURE NO. 4–6

ANGULARITY TOLERANCE, CYLINDRICAL SIZE FEATURE, RFS ∠

As discussed in the preceding section, a size tolerance does not control orientation of that feature. Where a size feature such as a hole or pin is shown on the drawing at an angle, (other than 90°), it must be controlled in some manner. As shown on page 54 under factors, a title block tolerance or standard may be invoked if there is no specified control of the angularity although this is sometimes questionable on a size feature. Does the implied tolerance apply to the surface or axis and under what condition (RFS, MMC or ?)? To clearly state the requirement on an angular feature, some specification is needed. The question is: What kind of control is required for the application?

DEVELOPING AN ANGULARITY TOLERANCE TO A SIZE FEATURE, RFS

First, if a hole or pin (a size feature) is shown on the drawing at an angle it will require some locational control. A positional tolerance will, of course, include an angularity tolerance. The angle, in that case, would be a basic angle.

Just as in preceding examples under perpendicularity of a size feature, an angularity tolerance, if applied, would be a refinement of the locational (i.e. position tolerance) control. Where needed, it is usually for control in only one direction as indicated by the view in which the requirement is specified. The amount of the angularity tolerance would be determined by the design requirement. Does it have to fit a pin inclined at an angle? If so, the angularity tolerance could be determined by the worst case (virtual conditions) of the mating features. Or, is it a hole which serves as a vent or for an inserted pin at assembly? Whatever its purpose, an angularity tolerance to a size feature requirement can be applied where necessary. The condition required, i.e., RFS or MMC, must be determined and stated.

AN ANGULARITY TOLERANCE TO A SIZE FEATURE PLACED ON THE DRAWING

The requirement should be shown in the view of the drawing where the angular relationship applies. Fig. 4-7 illustrates the part situation described above. Note the Basic angle 75°, the datum reference and the feature control frame and the manner in which they are placed on the drawing. The absence of the diameter symbol in the feature control frame indicates that the tolerance zone is a .005 total wide zone. The use of a cylindrical tolerance zone is not precluded. However, experience will show it is seldom needed. Let position tolerance control such things, if possible. If used, it presents some complicating factors to inspection.

The individual controlled feature must also be within the limits of size, boundary of perfect form at MMC (per Rule #1), and specified tolerance of location (latter not shown on this example).

MEANING AND VERIFICATION PRINCIPLES

As seen in Fig. 4-7, the tolerance is a .005 wide zone (distance between 2 parallel planes) inclined at 75° to the datum plane A. Regardless of feature size (RFS) is implied by Rule #3. Appropriate measuring tools would be applied using the principles illustrated.

ANGULARITY OF A CYLINDRICAL SIZE FEATURE

AS DRAWN

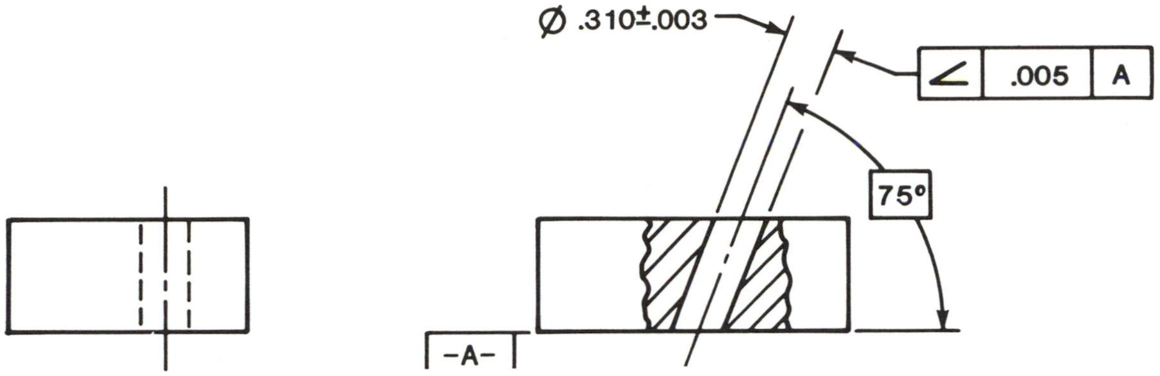

Ø .310±.003

∠ | .005 | A

75°

⌐-A-⌐

MEANING AND SAMPLE VERIFICATION PRINCIPLES

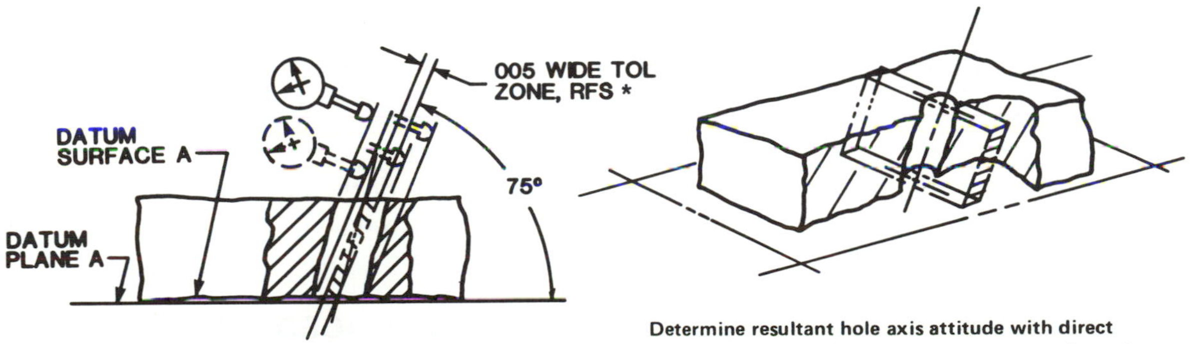

005 WIDE TOL ZONE, RFS *

DATUM SURFACE A

75°

DATUM PLANE A

Determine resultant hole axis attitude with direct measurements or differential measurements dependent upon hole basic form error (sloped or conical, etc.). Resultant axis must be within .005 wide tol zone, RFS. Also, hole must be within specified limits of size and location.

RFS CONDITION IMPLIED BY RULE #3.

IF METRIC

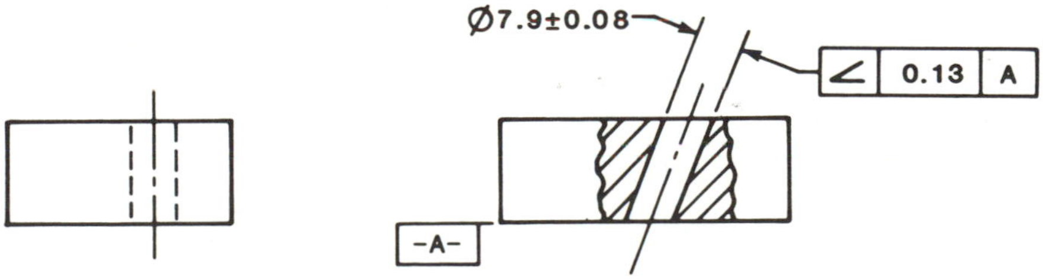

Ø7.9±0.08

∠ | 0.13 | A

⌐-A-⌐

FIGURE NO. 4–7

PARALLELISM TOLERANCE, CYLINDRICAL SIZE FEATURE, RFS //

As discussed in the preceding section, a size tolerance does not control orientation of that feature. Where a size feature such as a hole or pin is shown on the drawing as parallel to a surface (or another hole) but is not given an orientation tolerance, it will result as produced in manufacture. The accuracy of the produced feature, insofar as its parallelism, will be guided by the factors as listed on page 54 and the location tolerance specified.

DEVELOPING A PARALLELISM TOLERANCE TO A SIZE FEATURE, RFS

Suppose that the hole shown in Fig. 4-8 is relatively unimportant in its location (position) from the left face datum surface. But wherever it is, it is critical in its parallelism to the surface specified as datum A. The resulting parallelism of the hole (its axis) would be within the positional tolerance if nothing further had been stated on the drawing. That is, if the positional tolerance used datum A as its reference, parallelism would be automatically contained within, and as a part of, the positional control. However, if the hole must serve some design function which requires a specific parallelism as a refinement of that already achieved, a parallelism tolerance can be specified as shown at right. This requirement may be for a fit of some kind, for an inserted pin at assembly, etc., where its location can be lenient but its parallelism is important. Appropriate calculations and considerations would be made in determining the tolerance. The parallelism tolerance must be less than the position tolerance as a refinement. The position tolerance is not shown in this example to isolate the parallelism control. Later sections will cover position tolerance.

A determination also must be made as to whether the regardless of feature size (more precision) or the maximum material condition principles are required. In Fig. 4-8, RFS has been determined as the choice by the designer. Recalling RULE #3—The Other Than Position Tolerance Rule, RFS is automatically invoked in the feature control frame, as shown, if this is the desired condition.

In all orientation tolerances, a datum reference must be stated for parallelism. As described in preceding discussions, the parallelism tolerance applies to the axis of the hole. In the absence of the diameter symbol ($\varnothing$) in the feature control frame, the tolerance zone would be a total wide zone, (in this case a .005 wide zone).

A PARALLELISM TOLERANCE TO A SIZE FEATURE PLACED ON THE DRAWING

The feature control frame should be placed in a view, such as shown in Fig. 4-8, where the parallelism relationship is clearly seen with respect to its datum plane.

Note that the basic relationship of the controlled feature (the $\varnothing$.305 hole axis) is parallel to the datum plane A.

MEANING AND VERIFICATION PRINCIPLES

As is seen under Meaning in Fig. 4-8, the tolerance zone is .005 wide basically parallel to the datum plane. The axis of the produced hole must be within the .005 wide zone, RFS. In addition, the feature size tolerance must be met as well as the boundary of perfect form at MMC (Rule #1) and the specified tolerance of location (latter not shown on this example).

Sample Verification Principles illustrates the principles of inspection for the parallelism requirement using typical measuring tools. With the part stabilized to the datum plane, the resulting axis of the hole is derived (only in the parallel direction). If the hole is sloped, direct measurements can be used. If conical, differential measurements (difference between opposed element readings) must be used.

PARALLELISM, CYLINDRICAL SIZE FEATURE, RFS

AS DRAWN

-A-

$\emptyset .305 \begin{smallmatrix} +.003 \\ -.000 \end{smallmatrix}$

| // | .005 | A |

.XXX

MEANING

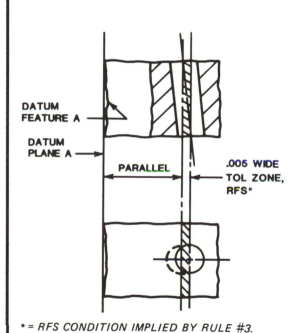

DATUM FEATURE A

DATUM PLANE A

PARALLEL

.005 WIDE TOL ZONE, RFS*

*= RFS CONDITION IMPLIED BY RULE #3.

IF METRIC

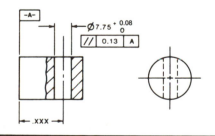

-A-

$\emptyset 7.75 \begin{smallmatrix} +0.08 \\ 0 \end{smallmatrix}$

| // | 0.13 | A |

.XXX

SAMPLE VERIFICATION PRINCIPLES

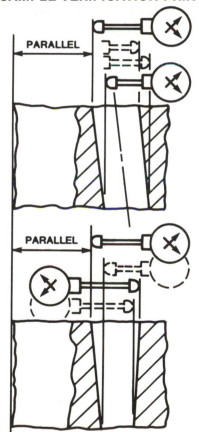

PARALLEL

PARALLEL

DETERMINE RESULTANT HOLE AXIS ATTITUDE WITH DIRECT MEASUREMENTS OR DIFFERENTIAL MEASURE-MENTS DEPENDENT UPON HOLE BASIC FORM ERROR (SLOPED OR CONICAL, ETC.). RESULTANT AXIS MUST BE WITHIN .003 WIDE TOL ZONE, RFS. ALSO, HOLE MUST BE WITHIN SPECIFIED LIMITS OF SIZE AND LOCATION.

FIGURE NO. 4–8

QUIZ-EXERCISES
ORIENTATION TOLERANCES

The following questions are relative to the material in this chapter. Read the question and answer to the best of your ability. The answers can be found in the companion manual *Answer Book and Instructor's Guide for Introduction to Geometric Dimensioning and Tolerancing.*

Consult your instructor if you have any questions.

1. Form, orientation, profile, and runout tolerances should be specified for all features critical to function and/or interchangeability and where:

   a. Established _____ practices cannot be relied upon to provide the required accuracy.

   b. Documents establishing suitable _____ of workmanship cannot be prescribed.

   c. Tolerances of _____ do not provide the necessary control.

2. The upper surface (upper extremity of the 1.610 dimension) of Figure 1 on page 85 is required in the part function to be in a parallel orientation of .002 total (maximum) tolerance to the surface. Add this requirement to Figure 1. Use letter A for the datum.

3. Show below (sketch) how the tolerance zone and the datum are established on the produced part (Figure 1) for the requirement of question 2. Assume the produced part surfaces are irregular.

4. Assuming the .002 parallelism tolerance, and the 1.610 + .005 size tolerance, what is the boundary of perfect form at MMC size? (Remember Rule #1.)

5. In Figure 1 (lower figure), suppose the vertical 1.610 surface is required to be in a square orientation to the lower surface within .003. Add this requirement to Figure 1.

6. Show below (sketch) how the tolerance zone is established for the requirement of question 5.

7. Suppose that in Figure 1 (upper view), the vertical 1.500 surface is required to be in a square orientation to the lower surface (of the 1.500 dimension) within .003. Add this requirement to Figure 1. Use letter B for the datum.

8. In questions 5 and 7 the 1.610 by 1.500 end face surface was controlled in its perpendicularity (squareness) in two directions from separate datums. Why are two separate specifications required? (Choose most significant reason from statements below).

   _____ Perpendicularity is preferably shown in the view most clearly depicting its relationship with its specific datum.

   _____ Perpendicularity controls form as well as orientation.

9. Assume that in Figure 1 the Ø.376 hole has been located with position dimensions and tolerance (do not yet concern yourself with the method), but the orientation of the Ø.376 hole must be maintained to a finer degree than the position tolerance. Specify on Figure 1 that this orientation control with respect to datum A is Ø.003 total, RFS.

10. Show below (sketch) how the tolerance zone is established for the requirement of question 9.

11. Referring to the perpendicularity tolerance used in question 10, what is the total tolerance permissible with hole size produced at $\varnothing.376$? _____. At $\varnothing.378$? _____.

12. The answers to question 11 are derived because: (Select the more correct answer.)

    _____ The tolerance is implied or stated as applicable RFS under Rule #3.
    _____ All tolerances stated are totals.

13. Suppose the perpendicularity tolerance of $\varnothing.003$ on the $\varnothing.376 \pm {}^{.002}_{.000}$ hole of Figure 1 (reference also question 9) was required by the part function to be on an MMC basis. How would the feature control frame be shown? (Show below.)

14. If the hole is produced at $\varnothing.376$ (MMC), what is the maximum permissible perpendicularity tolerance? _____

15. If the hole is produced at $\varnothing.378$, what is the maximum perpendicularity tolerance? _____

16. From questions 9 through 15 we see that whenever a feature of size such as a hole is involved, we must consider whether the conditions _____ or _____ are desired as a design requirement.

17. From the response to question 15, we see that use of the MMC principle, when appropriate to the design requirement, (gains/loses) _____ production tolerance yet assures function and interchangeability.

18. In Figure 1, 25° and 30° angles are critical to the extent of a .010 maximum tolerance variation as they relate to their respective datums A and B. Show these requirements on Figure 1.

19. Reviewing the Figure 1 questions and applications, it can be noted that of the four types of geometric characteristics used, three, _____, _____ and _____ require a datum reference.

20. _____ references are used wherever a specific relationship of one feature to another is required.

21. A datum plane is established from a _____ _____.

22. A datum _____ is an actual, physical, portion of a part.

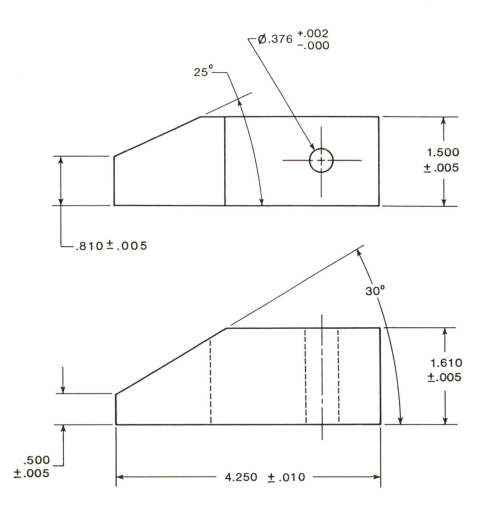

Figure 1

———————— **NOTES** ————————

—————————————————————————————————

—————————————————————————————————

—————————————————————————————————

—————————————————————————————————

—————————————————————————————————

—————————————————————————————————

5

PROFILE TOLERANCES

PROFILE OF A LINE ⌒ PROFILE OF A SURFACE ⌓

Profile tolerances include profile of a line and profile of a surface.

A profile of a line or a profile of a surface tolerance may be applied to individual features without a datum reference as a refinement of other controls in which case it serves as a special variety of form tolerance.

A profile of a line or a profile of a surface tolerance may be applied with a datum reference in which case it is a special variety of orientation tolerance.

Profile tolerance is a method used to specify a permissible deviation from the desired profile—usually an irregular shape where other geometric controls are inappropriate. A profile tolerance specifies a uniform boundary along the desired true profile within which the feature elements (surface or line) must lie. Profile tolerance is used to control form, or combinations of size, form, and orientation of a feature or features.

Profile tolerance may be applied to an entire surface of an individual profile taken at a cross-section or cutting-plane through a part or feature. The two methods are profile of a line and profile of a surface.

Datum references may, or may not, be used with profile tolerances. Usually, profile of a surface requires datums, whereas profile of a line may require datums less frequently.

Profile tolerances may be combined with other geometric tolerances. They may be used to refine other controls or may have other controls specified which refine within the profile control.

PROFILE TOLERANCES	INDIVIDUAL FEATURE	RELATED SURFACE FEATURES
Profile of Line ⌒　　**Profile of Surface** ⌓	No Datum reference is used.	A Datum feature is required.

87

PROFILE OF A LINE   ⌒

PROFILE OF A LINE TOLERANCE

The tolerance zone established by the profile of a line control is two dimensional extending along the length or width of the considered feature.

Profile of a line may be applied to any cross-section of a part including parts having varying cross-sections where it is necessary to describe such shapes, and/or where it is not desired to control the entire surface of the feature simultaneously.

The profile of a line tolerance is applied in a view of the drawing where the desired profile is seen and can be defined and described.

The profile of a line tolerance zone is a two dimensional zone and is the distance between two boundaries disposed about the desired true profile or entirely disposed on one side of the desired profile. The appropriate feature control frame, leader, and extent of the tolerance zone is shown to describe the type of zone desired.

Profile of a line tolerance applies normal to the line profile at all points along the line profile. The boundaries of the tolerance zone follow the shape of the true line profile.

Profile of a line tolerance is applied to parts to indicate shape of cross-sections and cutting-planes. Airframe, missiles, propellers, and impellers, etc., would represent typical parts where profile of a line could be used.

Profile of a line control is often a refinement of form within size control. It may also be a refinement of other geometric controls such as profile of a surface. In such cases, the profile of a line control is less than the control it refines.

Profile of a line control may require datum references where pertinent to design requirements.

See pages 90, 92, 93 under profile of a surface for additional detail on specifying a profile tolerance on the drawing. Similar methods and considerations may be applied to profile of a line.

DEVELOPING A PROFILE OF A LINE TOLERANCE

A profile of a line tolerance is appropriate where ''line elements'' only of a particular shape are required in the design. For example, imagine the part in Fig. 5-1 is a cam activator. The tangent R.380 profile and permissible profile tolerance will provide the ''throw'' (movement) required to transmit action to the concerned mechanism. This, of course, is determined by the design requirements. However, the follower which contacts (rides on) the profile surface will contact only at a point of tangency and move up or down on a ''line element'' on the surface. The illustrations under Meaning show the cutting planes, any one of which must meet the profile tolerance at the line element. In this case, the profile of a line control refines a size dimension (1.305 ± .015) and is also stabilized to two datum planes as specified (A and B). The third plane is unstated as it would be irrelevant to the requirement. This example utilizes a unilateral distribution of the tolerance. The Datum Reference Frame comprised of datums A and B (and the third unstated plane) will completely stabilize the part.

MEANING AND VERIFICATION PRINCIPLES

Production will derive a clear meaning from the drawing and will proceed with appropriate manufacturing tools and processes.

The drawings show how the part is stabilized to its datum planes A and B and how the measuring process principles will verify the part. Comparators or computer techniques may be used to advantage.

PROFILE OF A LINE

AS DRAWN

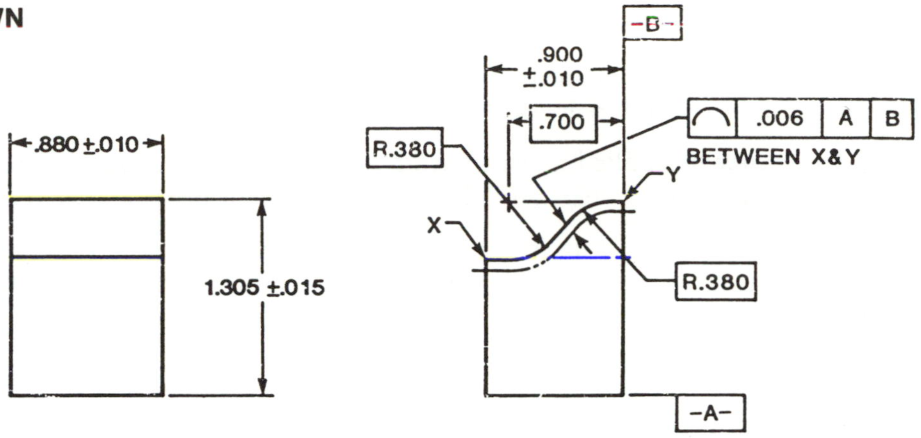

-B-

.900
±.010

.700

| ⌒ | .006 | A | B |

BETWEEN X & Y

R.380

R.380

.880 ±.010

1.305 ±.015

X

Y

-A-

MEANING AND VERIFICATION PRINCIPLES

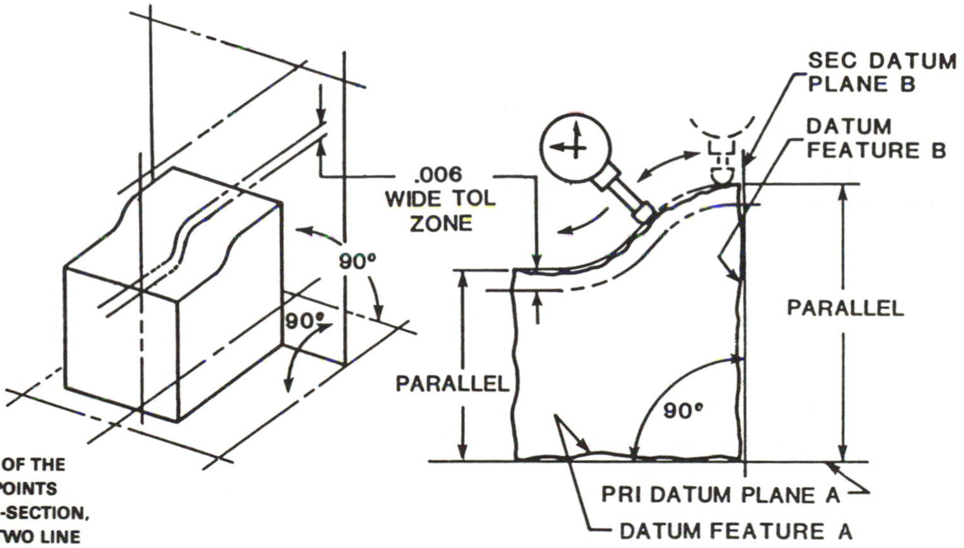

.006
WIDE TOL
ZONE

90°

90°

SEC DATUM
PLANE B

DATUM
FEATURE B

PARALLEL

PARALLEL

90°

PRI DATUM PLANE A

DATUM FEATURE A

EACH LINE ELEMENT OF THE
SURFACE BETWEEN POINTS
X & Y, AT ANY CROSS-SECTION,
MUST LIE BETWEEN TWO LINE
PROFILE BOUNDARIES .006
APART RELATIVE TO DATUMS
A AND B. ALSO, THE PROFILE
FEATURE (UNDER EXTREMITY)
MUST BE WITHIN THE SPECIFIED
SIZE LIMITS.

IF METRIC

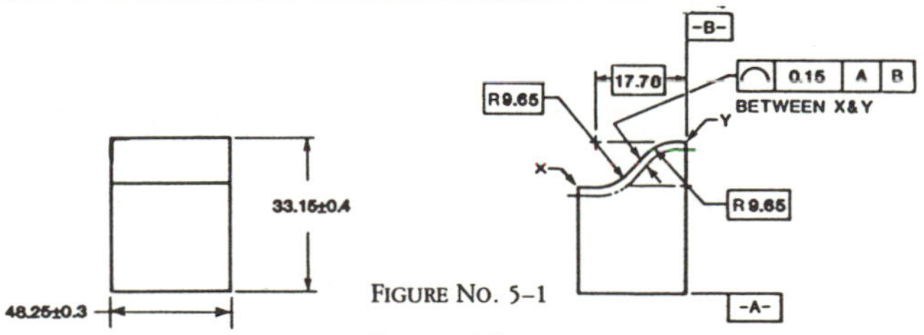

-B-

17.78

| ⌒ | 0.15 | A | B |

BETWEEN X & Y

R 9.65

R 9.65

33.15±0.4

48.25±0.3

X

Y

-A-

FIGURE NO. 5-1

FIGURE NO. 5-1

PROFILE OF A SURFACE

PROFILE OF A SURFACE TOLERANCE   ⌓

The tolerance zone established by the profile of a surface control is three dimensional extending along the length and width or circumference of the considered feature or features.

Profile of a surface may be applied to parts having a constant cross section or surfaces of revolution, or to those parts such as castings where an "all over" requirement may be desired.

A profile of a surface tolerance is a utility control which can provide a tolerance zone of any special shape (cams, arcs, radii, contours, fit, airframe, etc.). Profile of a surface tolerance can also be used on more common shapes (cones and co-planer surfaces, etc.). It can be applied to individual features, as a refinement of other controls, or with reference to datum features as the only control.

A profile of a surface tolerance can be applied as a variety of form tolerance, as a variety of orientation tolerance, or as a combination of size, form and orientation tolerances. A profile of a surface tolerance may be applied to any desired shape where appropriate to the design requirement. Occasionally the profile is established and defined by formulae or grid dimensions, etc.

When applying a profile of a surface tolerance to a part design, the profile needed is determined and is defined on the drawing as the true (exact) profile using Basic dimensions. The profile tolerance (zone) is determined as the permissible error of the surface relative to the true profile. The tolerance zone applies normal to the true profile.

DEVELOPING A PROFILE OF A SURFACE TOLERANCE—THE DESIGN REQUIREMENT

Suppose that the shape (profile) shown in Fig. 5-2 is the determined profile.

Let us assume that this shape must fit closely to a mating part shape and the surface relationships are important (see assembly at right).

The permissible tolerance deviation from that exact profile in the example is determined as a total of .010, .005 on either side of the true (exact) profile.

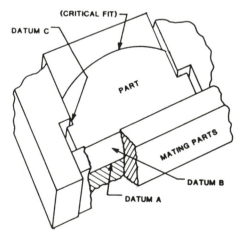

PLACING THE PROFILE OF A SURFACE ON THE DRAWING

A check list of considerations for establishing and properly specifying a profile tolerance follows:

    a. An appropriate view or section is shown in the drawing indicating the basic desired profile.

    b. The profile is defined by Basic (XXX) dimensions, angles, radii, arcs, etc.

PROFILE OF A SURFACE

AS DRAWN

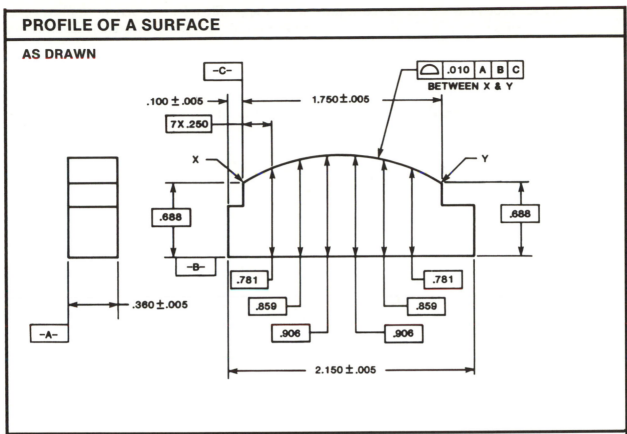

MEANING AND VERIFICATION PRINCIPLES

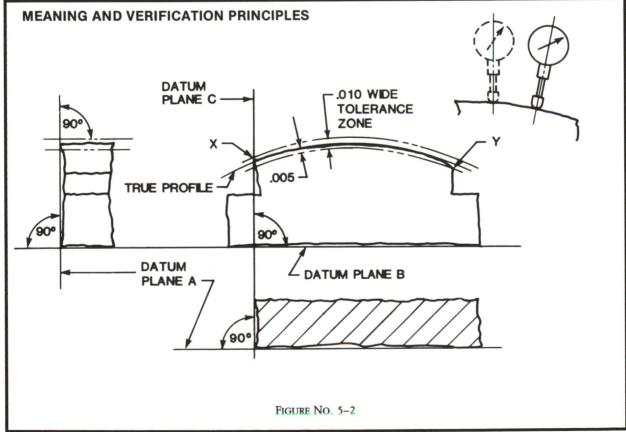

FIGURE NO. 5-2

c. Depending on the design requirements, the profile tolerance may be applied either bilaterally to both sides of the desired true profile or applied unilaterally to either side of the desired true profile (see below).

d. Any required datum references are determined.

In the example, it was determined that bilateral distribution of the tolerance relative to the true profile was desired. Therefore, the feature control leader extends directly to the profile on the drawing.

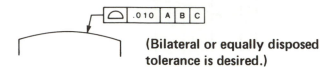

(Bilateral or equally disposed tolerance is desired.)

If, for some reason, it would have been more desirable to distribute the .010 tolerance unilaterally, one of the following methods would have been shown on the drawing to indicate all "outside" or all "inside" the true profile:

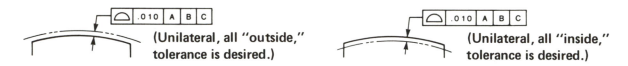

(Unilateral, all "outside," tolerance is desired.)

(Unilateral, all "inside," tolerance is desired.)

a. The phantom lines used to describe the tolerance zone need only to extend a representative distance to make its direction clear.

b. Where desired for clarity, letters or other suitable means may be used to identify the beginning and ending of such a tolerance zone. A brief note, such as "between X & Y," may be placed beneath the feature control frame to associate the elements of the requirement. This was considered desirable for clearer understanding in Fig. 5-2 although it is not required.

In this example, it was considered best to describe the "true" profile as a curve through selected calculated points (lines) using basic dimensions vertically and horizontally as derived from the radius curve. Since, in this case, the size of the part in the vertical direction is controlled by the profile tolerance, the vertical coordinates give some visible magnitude of the part size. The radius center would be in space off the part and, thus, would not easily suggest size if it were used to establish the basic curve. Such use would not be incorrect, however.

THE DATUM REFERENCE FRAME—DATUM PRECEDENCE

Note from the assembly drawing (page 90) and in Fig. 5-2 that the part rests (and stabilizes itself) on three surfaces at once. The nature of parts with more complex design requirements, more often than not, requires three datum references. This is called the Datum Reference Frame (also known as "The Three-Plane Concept"). It is the system of symbols which permits the part and its geometric tolerance requirement to be defined relative to the three planes, produced and verified in uniform compliance and assembled according to plan.

From the assembly drawing and in Fig. 5-2 we see that the design requirement rests the part on those three indicated datum features (surfaces) specified as A, B and C.

We must now extend the datum application from one datum feature, as discussed under orientation tolerances, to three datum features. The same principles involved with one plane surface datum now apply similarly to all three plane surfaces in this example. However, since the three surfaces (A,B,C) will not be perfect in their precision to one another (perfectly perpendicular) relative to the three 90° mutually related exact datum planes, an ''order of precedence'' is established. The most important functional surface is established as the primary datum, the next most important is called the secondary datum, and the third is called the tertiary datum. The letter in the feature control frame in the first compartment (left to right) is the primary datum (A in the example), the next compartment is the secondary datum (B) and the last compartment is the tertiary datum (C).

See also Pages 120, 121 and elsewhere under Position Tolerances in following text; see Supplemental Information section page 177, 178.

MEANING AND VERIFICATION PRINCIPLES

Production will derive a clear meaning of the design requirements and will proceed with suitable manufacturing processes. In Fig. 5-2 the relationship of the imperfect datum features to the datum planes A, B and C is shown. Note how the part is stabilized to the three planes for verification of the curved surface just as the part related to its mating parts in assembly. If dial indicators or hand methods are used, the indicator must register normal to the surface. Often optical comparator or computer assisted coordinate measuring machines (CMM) are used.

IF METRIC APPLICATION

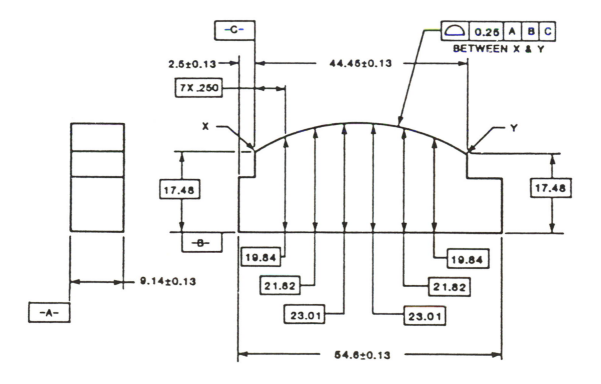

QUIZ-EXERCISES
PROFILE TOLERANCES

The following questions are relative to the material in this chapter. Read the question and answer to the best of your ability. The answers can be found in the companion manual *Answer Book and Instructor's Guide for Introduction to Geometric Dimensioning and Tolerancing*.

Consult your instructor if you have any questions.

1. Profile tolerancing is an effective method of controlling _____.

2. Where total surface control is desired, profile of a _____ control is used. Where line element control is desired, profile of a _____ control is used.

3. A profile tolerance is shown in the _____ of the drawing in which the desired profile appears.

4. The desired profile is dimensioned by _____ dimensions.

5. Profile of a surface control can be a combination _____ and _____ control.

6. Profile of a line control is normally used as a refinement of other _____ or _____ controls.

7. The surface profile on Figure 1 (next page) is to be controlled to a total of .010 equally disposed about the basic profile between X and Y and relative to datums A, B, and C. Show this on the drawing.

8. Show by hand sketch how the tolerance zone is determined in the preceding example.

9. Add to Figure 1 that the profile line elements shown in the front view are to be maintained to a finer tolerance (of .003) than the total surface profile.

10. The profile of any line tolerance zone must be controlled within the profile of any surface control shown in the answer to question 9. True or False _____.

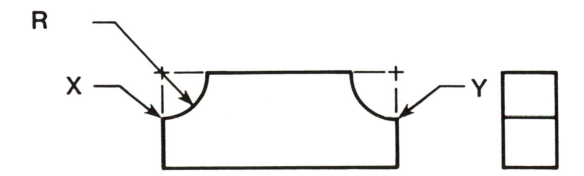

Figure 1

———————— **NOTES** ————————

_____

_____

_____

_____

_____

_____

6

RUNOUT TOLERANCES

CIRCULAR RUNOUT   ↗   TOTAL RUNOUT ⤴

Runout tolerances include circular runout and total runout.

Circular runout provides control of circular elements of a surface. Total runout provides composite control of all surface elements. Runout tolerance may be applied to surfaces such as cylinders, cones, and other surfaces of revolution around a datum axis or to surfaces constructed at right angles to a datum axis.

Runout tolerance always requires a datum axis. Where the datum is an outside cylindrical feature, the simulated datum cylinder and its derived datum axis are established by the minimum circumscribed cylinder which will contact (close upon) the extremities of the datum feature. A bearing mount to the datum feature would be a typical design requirement (represented by a precision collet in inspection). See pages 108, 109 for details on Datums and the Datum Reference Frame.

Runout tolerance is applicable only on an RFS basis. See also Supplemental Information, page 188.

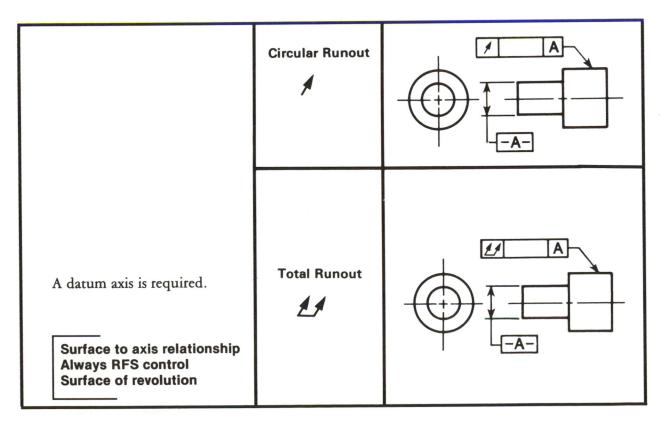

A datum axis is required.

Surface to axis relationship
Always RFS control
Surface of revolution

Circular Runout

↗

Total Runout

⤴

COAXIAL FEATURES

There are three types of coaxial feature control—runout, position, and concentricity. Proper selection is based upon which best suits the functional design requirement.

Runout—Use runout where part feature surfaces in a rotational consideration must relate to a datum axis. Runout is applicable only on an RFS basis. Typical callout:

CIRCULAR RUNOUT | ↗ | .003 | A |

TOTAL RUNOUT   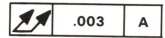 | .003 | A |

(surface-to-axis control, RFS)

Position—Use position where part feature surfaces relate to a datum axis on a functional or interchangeable basis. Typically, mating parts are involved. Position is normally applied only on an MMC basis. Occasionally an RFS datum is used. Typical callout:

| ⊕ | ⌀ .003 Ⓜ | A Ⓜ |

(surface-to-surface control, MMC)

(axis-to-axis control, MMC)

Concentricity—Concentricity is used where part feature axis (or axes) in a rotational consideration must relate to a datum axis. Concentricity is applicable only on an RFS basis. Typical callout:

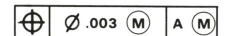

(axis-to-axis control, RFS)

DEFINITIONS

Runout is the composite deviation from the desired form of a part surface of revolution during full rotation (360°) of the part on a datum axis. Runout tolerance may be circular or total.

Circular runout is the composite control of circular elements of a surface independently at any circular measuring position as the part is rotated through 360°.

Total runout is the simultaneous composite control of all elements of a surface at all circular and profile measuring positions as the part is rotated through 360°.

RUNOUT TOLERANCE

A runout tolerance is applied to control precision of surfaces of revolution. Rotating parts such as drive shafts, cam shafts, spindles, etc. would be typical. However, any such part upon which a datum axis is established could have runout tolerance applied as appropriate to the design requirement.

Runout is a relationship type control which can be used to place limits (a tolerance) upon the composite effect of surface error of a feature about the datum axis. Circular runout, for example, is a composite of the effects of out-of-circularity and concentricity. Total runout is a composite of the effects of out-of-circularity, straightness, parallelism of opposed elements (taper), cylindricity, and concentricity. Runout can also be applied to control composite error of surfaces at 90° to a datum axis, or any symmetrical surface which is about a datum axis.

Since it is a precision control, runout is applied only on an RFS basis.

DEVELOPING A RUNOUT TOLERANCE-DESIGN REQUIREMENT

Typical use of runout tolerance would be on a bearing mounted shaft upon which components such as gears, pulleys, sprockets, clutches, etc. are mounted. The amount of composite surface error (tolerance) which can be permitted to be transmitted from the shaft to the component is determined by the designer. Runout can be applied directly from the calculated tolerance to the drawing. Usually, the designer is striving for a composite control. The calculations usually determine the composite amount of tolerance and runout tolerance provides a method of stating the requirement as a composite value. The type of runout (circular or total) is a design decision which will determine the level of precision stated. Therefore, there is a need to have a clear distinction between these two options. A comparison is made in Fig. 6-1.

Suppose that the part shown mounts to a bearing on the datum A feature and thus establishes the datum A axis of rotation. Imagine that a gear mounts on the larger diameter. The gear mounts on the surface of the diameter yet it will rotate about the datum axis. A decision is then required as to which coaxial feature control of three types available (runout, position or concentricity) is most appropriate. Runout tolerance is usually the best for such a precision application as shown where surface error on an RFS basis is desired. Position tolerance, using the MMC principle, is possible but not often desirable in this type of application. Concentricity is also possible and could be used in this situation. However, as will be discussed later, it could permit less surface precision in some cases and yet be acceptable to a concentricity tolerance at the axis of the feature controlled.

CIRCULAR RUNOUT APPLIED

Referring to Fig. 6-1, suppose the permissible composite surface error on the large mounting diameter is .002 as developed using the above reasoning. A circular runout callout is included in the feature control frame and the datum is specified as shown. The datum feature is specified but the datum is the axis of the datum feature.

MEANING AND VERIFICATION PRINCIPLES

Circular runout principles are shown in Fig. 6-1. An infinite number of circular elements are involved and any individual circular element checked, as shown, must be within .002 FIM as the part is rotated 360°. A typical measuring tool, a dial indicator, is used as representative of the measuring process. Each circular element checked must individually meet the circular runout requirement. An inspector will usually sample only cross-sectional places and not check every element. Note that the surface error could permit some taper (and other possible errors within the size tolerance) and yet possibly meet the circular runout tolerance. It is a two-dimensional control, applied only at cross-sections of the part. Such a control could be very adequate, however, for the application and also be more economical in production. The decision for the kind of control necessary rests with the designer.

TOTAL RUNOUT APPLIED

Fig. 6-1 also shows a total runout version of the same part as under circular runout. In this case the designer, using determinations discussed earlier, has decided to use a more stringent control. Total runout is a three-dimensional control and will account for the entire (total) surface in one tolerance (.002). The taper (and other possible errors) are not prevented, but are controlled as a composite over the entire length of feature in one tolerance zone. This is a higher precision part and, thus, probably more costly. The designer makes the decision based upon the design requirements.

MEANING AND VERIFICATION PRINCIPLES

Comparing the principles of both of the above examples in Fig. 6-1 should clarify the major differences between circular and total runout. As seen, in the inspection process, total runout requires that the .002 FIM be met collectively (*one setting* of the indicator) as the inspector rotates the part through 360° at various places on the controlled diameter.

In both examples the parts are held on the datum feature by a collet (or chuck) as a representative inspection device. Other methods of inspection are, of course, available but would utilize the same principles. This is a "simulation" of the bearing mount.

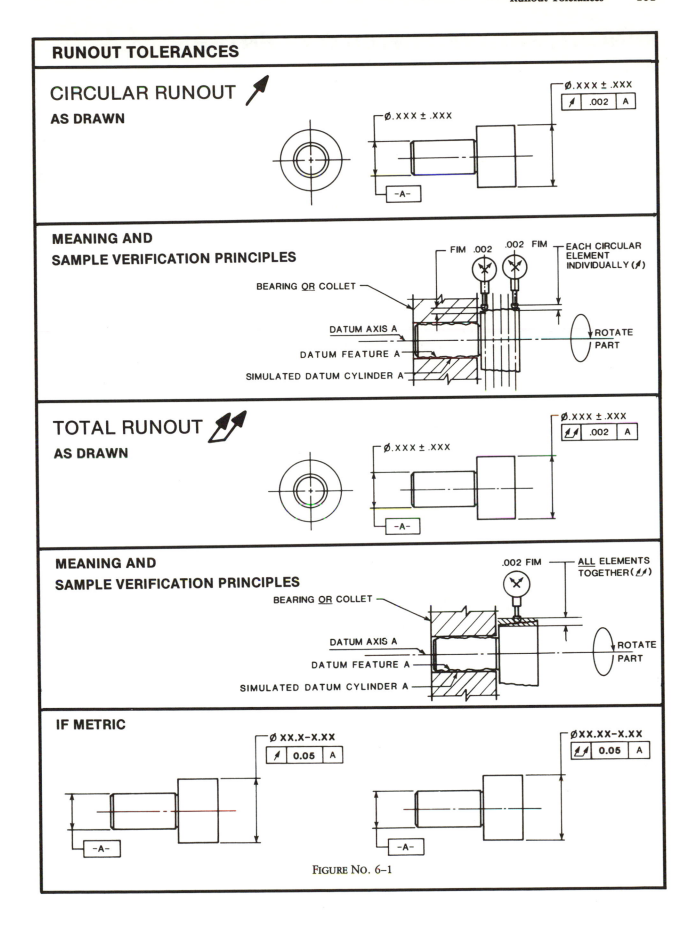

RUNOUT TOLERANCES

CIRCULAR RUNOUT

AS DRAWN

MEANING AND
SAMPLE VERIFICATION PRINCIPLES

TOTAL RUNOUT

AS DRAWN

MEANING AND
SAMPLE VERIFICATION PRINCIPLES

IF METRIC

FIGURE NO. 6-1

RUNOUT, PART MOUNTED ON CENTERS

COMPOUND (MULTIPLE) DATUMS

Runout tolerance can be related to a datum axis established by two part centers as shown in Fig. 6-2. In such a case, the part function and final mounting is on the centers. The two centers are individually specified as datum features thus establishing a common datum axis about which the part rotates. This is called compound or multiple datums. The centers themselves are identified as the datums, not the drawing centerline. The multiple datum callout (A-B) is placed in the feature control frame as the datum reference. There is no precedence between the two datum features. They function together to establish the common datum axis.

THE DESIGN REQUIREMENT

Imagine that this part mounts on two live centers in assembly on the datum features A and B. The runout tolerance for each diameter of the step-shaft is calculated based upon the permissible error that each diameter can transmit to the component to be mounted as it rotates. The designer determined that this application required total runout control due to the precision of the part function. Each calculation can be placed directly on the part drawing as a runout tolerance. This avoids constructing such a control with numerous elements of other controls (circularity, straightness, concentricity, etc.) or explanatory notes which should be avoided.

MEANING AND VERIFICATION PRINCIPLES

Production prefers composite requirements, such as runout, on the drawing rather than a number of single requirements to achieve the same thing. It is easier to understand and it is the manner in which such parts are normally machined.

Where the datum axis is established from two internal centers, it is the maximum inscribed true cone which will contact the extremities of the two produced centers (cones). The two true cones can be represented by machining or inspection centers. The two true centers are assumed to be collinear and precise with tool or gaging tolerance.

The features related to the datum axis in a runout specification must be within the stated runout tolerance (FIM) when rotated 360° about the datum axis. In addition, such features must be within their respective specified size limits and within the boundary of perfect form at MMC of each individual feature.

Where total runout is specified, the concerned surface must be within the stated runout tolerance across the entire (total) feature (FIM) collectively when rotated 360° about the datum axis.

The part function, its manufacture, and verification are uniformly stated and communicated with runout tolerance. The designer may directly state the tolerance as the composite result of design calculations, and production can tool and machine more directly in accordance with the composite tolerance given. Verification is accomplished in one set-up with minimum inspection steps.

RUNOUT, PART MOUNTED ON CENTERS—MULTIPLE DATUMS

AS DRAWN

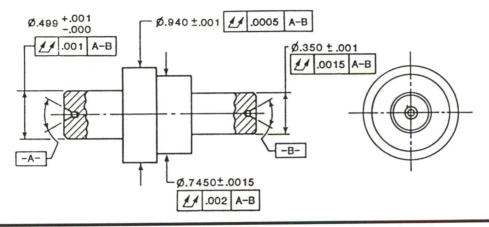

MEANING AND SAMPLE VERIFICATION PRINCIPLES

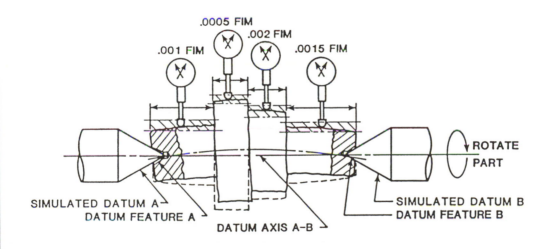

⟋⟋ = ALL ELEMENTS VERIFIED TOGETHER (TOTAL).

IF METRIC

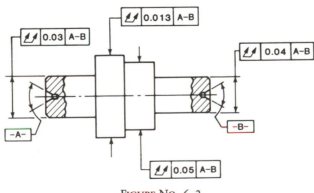

FIGURE NO. 6–2

RUNOUT, PART MOUNTED ON TWO FUNCTIONAL DIAMETERS

COMPOUND (MULTIPLE) DATUMS

Runout tolerance is ideally suited to parts which mount to bearings. The bearing mount diameters (cylinders) are selected as the datum features and identified (A and B) with the related features given the calculated runout tolerances. Each runout tolerance is specified with an appropriate feature control frame including the datum references to which the concerned feature is related. Each feature controlled by a runout tolerance must be within the stated tolerance (FIM) when the part is rotated 360° about the datum axis. The datum axis (A-B) is the axis established by two collinear minimum circumscribed cylinders which contact (close on) the datum features simultaneously. This represents very closely the part features mount to the bearings. Such datums are called compound or multiple datums. Where the selected datum features themselves are not given geometric controls, they are assumed to establish the datum axis from their "as produced" conditions and relationship. See Figure 6-3.

Where accumulation of runout error may need to be controlled between specific features, one of the features may be selected and specified as a datum feature. The second feature is then related to it with an appropriate feature control frame and referenced to that datum. In such cases, the maximum runout error permitted between the two features is that stated.

THE DESIGN REQUIREMENT

Runout tolerance, when applied to a step shaft where components such as gears, pulleys, sprockets, cams, clutches, etc. are to be mounted, can provide the designer a more direct means of stating composite permissible error (tolerances). Rather than determine individual permissible tolerances in design calculations for construction of a composite result, it can be done directly with runout tolerances. Further, the functional relationship of the features to the datums (simulating the bearings) is direct and representative.

Runout can also be applied to surfaces at right angles to the datum axis. Such application can control perpendicularity, wobble, etc. in composite relative to the functional datum axis. Runout tolerance of this kind eliminates variations in interpretation which can accompany a perpendicularity specification, for example, on such an application; does the designer wish to indicate that the part is fixed (not rotating) when verified, or rotated, one shoulder at a time, or total error across both shoulders of the part, etc.). Runout eliminates such a communication problem. The part must rotate upon verification (same as part function) and represents total permissible error in such rotation about the datum axis.

MEANING AND VERIFICATION PRINCIPLES

Production follow-through is also assisted as the composite tolerances blend directly with tooling and machining or manufacturing methods and does not require numerous individual operations to meet numerous individual requirements. Verification procedures in such an application are condensed to minimum set-ups (two in this case) and conventional techniques can be used. Fig. 6-3 clearly shows the process.

All features involved in a runout tolerance, including the datum features, must individually be within their respective stated limits of size and perfect form at MMC boundaries.

RUNOUT, PART MOUNTED ON TWO FUNCTIONAL DIAMETERS

AS DRAWN

Ø .605 ± .003
⟋⟋ .002 A–B

⟋⟋ .002 A–B

⟋ .001 A–B

Ø .750 ± .001
–A–

Ø .375 ± .002
⟋⟋ .001 A–B

Ø .8800 ± .0005
⟋⟋ .001 A–B
–R–

Ø 1.200 ± .003

Ø .810 ± .001
–B–

Ø .500 ± .001
⟋⟋ .0005 R

MEANING AND VERIFICATION PRINCIPLES

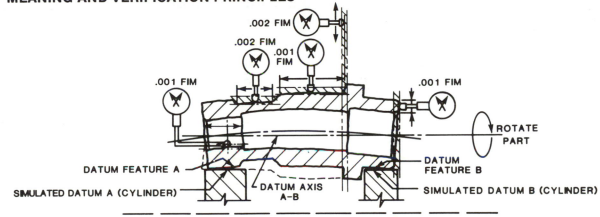

.002 FIM

.002 FIM

.001 FIM

.001 FIM

.001 FIM

ROTATE PART

DATUM FEATURE A

SIMULATED DATUM A (CYLINDER)

DATUM AXIS A–B

DATUM FEATURE B

SIMULATED DATUM B (CYLINDER)

⟋⟋ = ALL ELEMENTS VERIFIED TOGETHER (TOTAL)

MEANING AND VERIFICATION PRINCIPLES

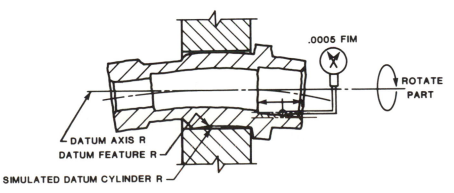

.0005 FIM

ROTATE PART

DATUM AXIS R
DATUM FEATURE R
SIMULATED DATUM CYLINDER R

⟋⟋ = ALL ELEMENTS VERIFIED TOGETHER (TOTAL)

FIGURE NO. 6–3

RUNOUT, PART MOUNTED ON FUNCTIONAL FACE AND DIAMETER

PRIMARY AND SECONDARY DATUMS

Runout tolerance can be applied to parts where a surface perpendicular to the axis and the axis itself are used as datums. In such cases, selection of the primary datum, and possibly a secondary datum, while considering the need for datum precedence, is usually necessary. The part function will determine whether the surface perpendicular to the axis is most critical or the axis is most critical. Thus, establishment of the primary datum will be determined.

Where the function determines that a face perpendicular to the axis of rotation is most critical, it is established as the primary datum. The secondary datum is established as the diameter (usually the mounting diameter or cylinder to a bearing) which determines the axis of rotation. The face datum is established by a plane in a contact with the extremities of the produced surface. The secondary datum cylinder and, thus the datum axis, is established by the minimum circumscribed (simulated datum) cylinder, at 90° to the primary datum plane, in contact with the produced cylinder. See Figure 6-4.

Where total runout is used, composite control of the entire (total) rotating surface is achieved. This includes the collective effect of errors of circularity and concentricity while the concerned surface rotates about the datum axis and is squared-up to the datum face.

Each datum reference and each runout tolerance is specified with an appropriate datum and feature control frame. Datum precedence is established by the primary and secondary datum reference letters, reading left to right, in the feature control symbol (A,B).

Where the composite effect of a rotating surface is to be controlled, as runout provides, circular runout should be considered first. If satisfactory to the design requirements, circular runout can usually be more economically produced and verified. Total runout is applied more normally to cylindrical or face mounting and functional square-to-the-axis surfaces.

Where runout is applied in the square-to and datum axis relationship, more unique design requirements can be readily stated. Uniformity of understanding and compliance in production and verification can also be more readily achieved.

MEANING AND VERIFICATION PRINCIPLES

Clarity of the drawing requirements will assist production and, thus, manufacture of the part. Inspection will be guided by the specific meaning of the feature control frame call-out. Fig. 6-4 shows the principles involved.

Where a measuring device such as a dial indicator is used, it is mounted normal to the desired profile (the concerned surface) and must be within the stated runout tolerance (FIM) when the part is rotated 360° about the datum axis.

All individual features involved in a runout requirement must be within the specified limits of size and the boundary of perfect form at MMC.

RUNOUT, PART MOUNTED ON FUNCTIONAL FACE AND DIAMETER

AS DRAWN

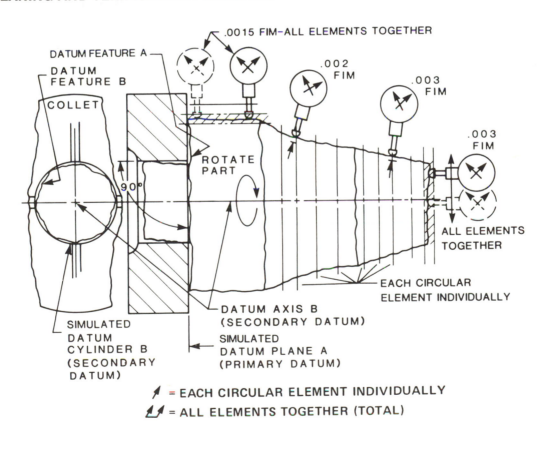

| ⌀⌀ | .0015 | A | B |

| ⌀ | .002 | A | B |

| ⌀ | .003 | A | B |

| ⌀⌀ | .003 | A | B |

−A−

−B−

MEANING AND VERIFICATION PRINCIPLES

.0015 FIM—ALL ELEMENTS TOGETHER

DATUM FEATURE A

DATUM FEATURE B

COLLET

.002 FIM

.003 FIM

.003 FIM

ROTATE PART

90°

ALL ELEMENTS TOGETHER

EACH CIRCULAR ELEMENT INDIVIDUALLY

SIMULATED DATUM CYLINDER B (SECONDARY DATUM)

DATUM AXIS B (SECONDARY DATUM)

SIMULATED DATUM PLANE A (PRIMARY DATUM)

⌀ = EACH CIRCULAR ELEMENT INDIVIDUALLY

⌀⌀ = ALL ELEMENTS TOGETHER (TOTAL)

FIGURE NO. 6–4

DATUMS—DATUM REFERENCE FRAME THREE PLANE CONCEPT `-A-`

DEFINITION

A datum is a theoretically exact point, axis, or plane derived from the true geometric counterpart of a specified datum feature. A datum is the origin from which the location or geometric characteristics of features of a part are established.

A datum feature is an actual feature of a part which is used to establish a datum.

BASIS FOR THE DATUM REFERENCE FRAME— CYLINDRICAL AND NON-CYLINDRICAL FEATURE

Where a size feature of a part such as a cylinder or width is used as a datum on an RFS basis, the derived datum axis or centerplane is established by the contact on the part feature with a simulated datum counterpart cylinder (the minimum circumscribed cylinder—such as a precision collet op chuck—which can be closed upon the datum feature, cylinder, etc.) or opposed parallel planes. See Fig. 6-5 and other appropriate sections of this text for further explanation.

Use of the datums on an RFS basis or MMC basis is, of course, a choice made by the designer. Whether the datum is referenced on an RFS basis for greater precision and more stringent control or on an MMC basis as a more functional interface of mating parts, etc., is based upon design requirements.

DATUM ESTABLISHMENT · RFS

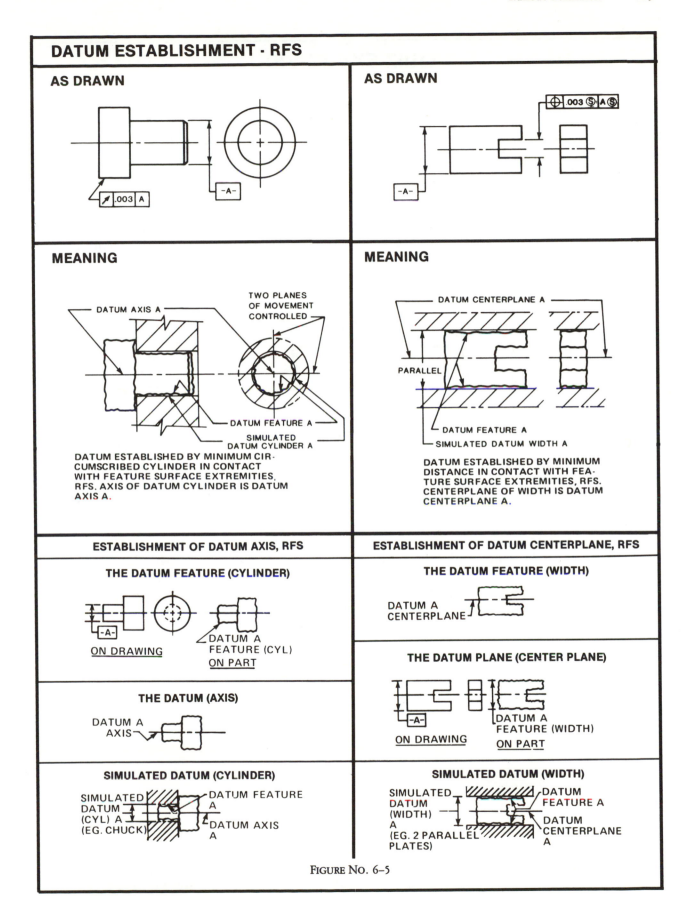

AS DRAWN

AS DRAWN

.003 Ⓢ A Ⓢ

-A-

-A-

MEANING

DATUM AXIS A

TWO PLANES
OF MOVEMENT
CONTROLLED

DATUM FEATURE A

SIMULATED
DATUM CYLINDER A

DATUM ESTABLISHED BY MINIMUM CIR-
CUMSCRIBED CYLINDER IN CONTACT
WITH FEATURE SURFACE EXTREMITIES,
RFS. AXIS OF DATUM CYLINDER IS DATUM
AXIS A.

MEANING

DATUM CENTERPLANE A

PARALLEL

DATUM FEATURE A

SIMULATED DATUM WIDTH A

DATUM ESTABLISHED BY MINIMUM
DISTANCE IN CONTACT WITH FEA-
TURE SURFACE EXTREMITIES, RFS.
CENTERPLANE OF WIDTH IS DATUM
CENTERPLANE A.

ESTABLISHMENT OF DATUM AXIS, RFS

THE DATUM FEATURE (CYLINDER)

-A-

ON DRAWING

DATUM A
FEATURE (CYL)
ON PART

THE DATUM (AXIS)

DATUM A
AXIS

SIMULATED DATUM (CYLINDER)

SIMULATED
DATUM
(CYL) A
(EG. CHUCK)

DATUM FEATURE
A

DATUM AXIS
A

ESTABLISHMENT OF DATUM CENTERPLANE, RFS

THE DATUM FEATURE (WIDTH)

DATUM A
CENTERPLANE

THE DATUM PLANE (CENTER PLANE)

-A-

ON DRAWING

DATUM A
FEATURE (WIDTH)
ON PART

SIMULATED DATUM (WIDTH)

SIMULATED
DATUM
(WIDTH)
A
(EG. 2 PARALLEL
PLATES)

DATUM
FEATURE A

DATUM
CENTERPLANE
A

FIGURE NO. 6-5

The following questions are relative to the material in this chapter. Read the question and answer to the best of your ability. The answers can be found in the companion manual *Answer Book and Instructor's Guide for Introduction to Geometric Dimensioning and Tolerancing.*

Consult your instructor if you have any questions.

1. Referring to Figure 1, assume that the part diameters on each end are to mount into bearings and the other diameters are to be within .002 total (FIM) relative to the part axis of rotation regardless of feature size. _____ tolerancing should be used. Show the requirements on Figure 1.

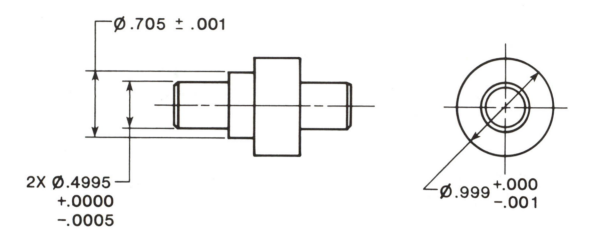

Figure 1

2. A runout tolerance relates (rotational/nonrotational) _____ surfaces of a part to a _____ axis. These surfaces may be (central/coaxial) _____ with the datum axis or (perpendicular/radial) _____ to the datum axis. The condition (RFS/MMC) _____ is always used in runout tolerancing.

3. A runout tolerance establishes a means of controlling the functional relationship of two or more features of a part. This type of tolerance is a (complex/composite) _____ type and may be applied as one of two different types of runout. These two types are _____ runout and _____ runout.

   a. Total runout includes such form and orientation errors as _____
   _____.

   b. Circular runout includes such errors as _____, _____ and circular _____ of the surface when applied to surfaces constructed at right angles to a datum axis.

4. Runout tolerance is considered as a unique category of geometrical tolerances and is actually a combination of _____, _____ and _____ controls in composite.

5. A datum axis for a runout tolerance may be established by a diameter (cylinder) of considerable length, two diameters having axial separation, or a diameter and a _____ which is at a _____ angle to it.

6. Referring to Figure 2, assume that the left face of the 1.0300 diameter is to be the primary datum and the .4995 diameter is to be the secondary datum and provides the axis of rotation. The part mounts into a bearing. Specify the datums with proper precedence and the .890 and 1.0300 diameters within total runout of .001 with respect to the datum axis.

7. Assume that the circular elements of the 45° angular surface of Figure 2 are required to run true in rotation within .0005 total. Add this requirement to Figure 2.

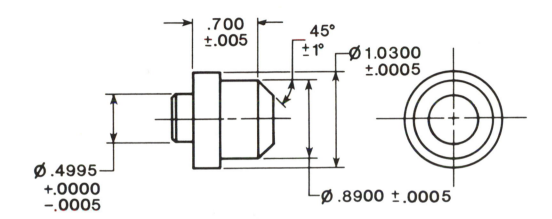

Figure 2

NOTES

7

LOCATION TOLERANCES

POSITION ⊕   CONCENTRICITY ◎

Location tolerances include tolerances of position and concentricity.

Location tolerances always deal with size features and are used to control the following types of relationships:

center distances between features such as holes, pins, slots, projections, etc.,

location of features as a group relative to a datum, or datums,

coaxiality between a feature, or features, relative to a datum axis, and

centrality between a non-cylindrical feature, or features, relative to a datum centerplane.

Since tolerances of location deal with size features, appropriate consideration to the desired principles of RFS or MMC must be made.

Location tolerances relate to specified datum references. The Datum Reference Frame (three plane concept) datum precedence, and multiple datums are involved in location tolerance considerations.

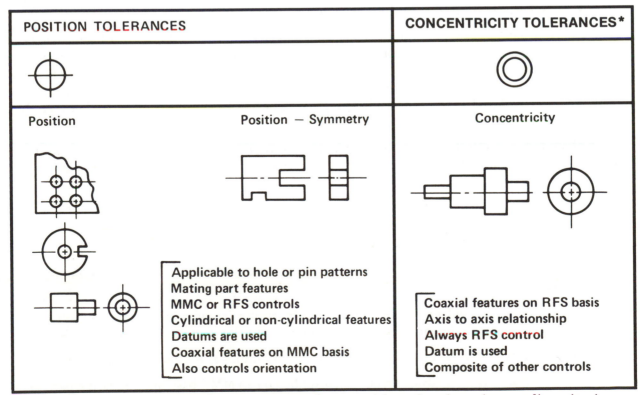

POSITION TOLERANCES	CONCENTRICITY TOLERANCES*
⊕	◎
Position Position — Symmetry	Concentricity
Applicable to hole or pin patterns Mating part features MMC or RFS controls Cylindrical or non-cylindrical features Datums are used Coaxial features on MMC basis Also controls orientation	Coaxial features on RFS basis Axis to axis relationship Always RFS control Datum is used Composite of other controls

*NOTE: See Pages 168 and 169 for text coverage of concentricity as location tolerance discussion is continued in Chapter 7.

POSITION TOLERANCE THEORY   ⊕

Position tolerance, traditionally called true position tolerance in the past, has the most potential advantages and is the most important segment of geometric dimensioning and tolerancing. The entire system of geometrics is sometimes referred to as the ''True Positioning System.''

Position tolerance provides the ideal method to ensure proper mating of parts via hole patterns, pin patterns, and almost any kind of mating feature locational situation. In Chapter 1 the type of part interface in which position tolerance can assist is discussed and illustrated. The parts and features shown in that chapter provide a typical situation where function and relationship is involved and can be specified with position tolerancing. Datum references are always used with position tolerancing as shown on the Flange Mount part in Chapter 1. A review of this chapter is suggested here as a preliminary to further study of position tolerance.

Position tolerance methods provide the ability to easily calculate tolerances and ensure 100% interchangeability of the concerned features. It captures the ''personality'' of the part and, along with the use of the appropriate material condition symbols (modifiers Ⓜ, Ⓢ, Ⓛ,) can represent the part function and relationship as no other method can.

The Datum Reference Frame is fully utilizied in relating the holes, pins, etc. as related to the appropriate datums. In addition, the MMC concept can be implemented to gain ensured part interchangeability, increased production tolerances, universal interpretation and functional gaging if desired.

Position tolerance incorporates most of the elements of the system already discussed in earlier sections on limits of size, form, orientation, profile and datums. On pages 120 and 121 the Datum Reference Frame is expanded and explained.

The earlier discussion on profile tolerance introduced the datum reference frame relationship which now becomes fully utilized in position tolerancing.

Because of the many uses of position tolerance, the material condition under which the tolerance is to apply must be stated in the feature control frame. Rule #2—The Position Tolerance Rule, one of the governing principles of the USA standard, requires this to be done. If MMC principles can be used, the greatest all around advantages are achieved from design to production. In the example following, the MMC ppinciple is used to demonstrate its advantage and to best distinguish the difference between the ± and position systems.

Perhaps the most effective way to become familiar with the advantages and principles of position tolerance is to compare it with the coordinate system as referenced in Chapter 1, Fig. 1-3 and as further discussed following.

ADVANTAGES OF POSITION TOLERANCE

Position tolerance is one of the most effective and most used controls in geometric dimensioning and tolerancing.

Position tolerance provides the ability to realistically capture design requirements, simplify calculations, ensure interchangeability, and use most economic production and verification processes.

Position tolerance may be used in all the feature relationships as indicated on page 113.

Position tolerance is most advantageous when applied on an MMC basis and in mating part relationships. It provides techniques and methodology which recognize and capitalize on the manner in which part features, such as holes and pins, relate to one another in assembly.

COMPARISON BETWEEN COORDINATE AND POSITION TOLERANCE METHODS

Coordinate tolerancing methods do not capture the true feature interface between mating parts. Mating holes and pins, for example, usually relate to one another in 360° of movement, yet the coordinate ± system cannot specify such a tolerance. Although correct design consideration may be made between the hole and mating pin in calculations or layout, the true relationship and interplay of size and location is lost in attempting to state such requirements via the coordinate system. See below and following exploratory discussion to rationalize the reasoning when comparing the ± system and the position tolerance system.

DESIGN REQUIREMENT

Part with holes and part with pins to mate-up at assembly (using the coordinate ± method).

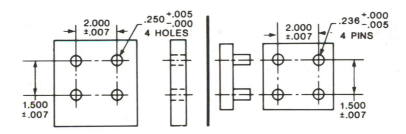

If the part with holes is shown with equally distributed resulting tolerance zones, the analysis below can be made. This will reveal some of the fallacies and shortcomings of the coordinate system as it compares to the position system.

STEP 1

DEDUCTION: With pins and holes of mating parts in perfect desired location, clearance at assembly is seen as uniform all around (Fig. 1) (with low limit size hole, high limit size pin). If holes and pins are produced at opposite limits of coordinate tolerance in X direction (Fig. 2) or in Y direction (Figs. 3 & 4), i.e., ± dimensioning, a square tolerance zone results if each hole is allocated its equal share of the coordinate tolerance in X and Y. (Fig. 5).

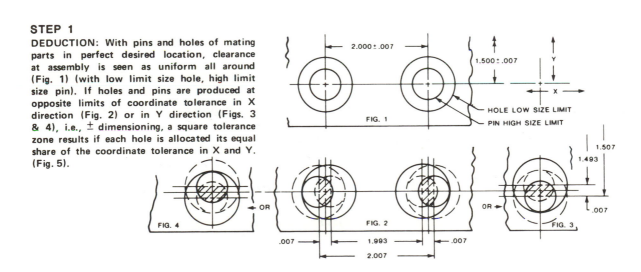

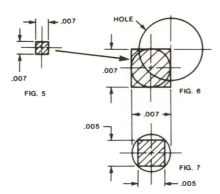

FIG. 5

FIG. 6

FIG. 7

STEP 2

NEXT DEDUCTION: Examination reveals that the .007 square tolerance zone could permit a hole to be produced *within* the zone but actually cause interference at assembly; the .007 tolerance can be permitted *only* on the X and Y axis, *not* diagonally. i.e., a .007 diameter tolerance zone. (Fig. 6)

STEP 3

IF, the ∅.007 tolerance zone is reduced (inscribed square) by calculation (.707 x .007 = .005) to safely establish an X and Y tolerance zone where hole center production will ensure assembly, production tolerance is reduced 36%. However, .005 is the maximum tolerance which can be allowed to describe the total tolerance zone in X and Y. (Fig. 7)

Position tolerance methods can directly utilize the diametrical (cylindrical) tolerance zone, permit direct calculations, and capture the true design requirements. In addition, the MMC principle can be used which is not possible with the ± system.

Fig. 7-1 compares the part using the coordinate system (upper), with the positional system (lower). The black dots indicate possible produced centers of one of the holes in five parts. The hole center on the upper left diagonal corner of the square zone at mid-page is the only acceptable hole to both systems. The hole on the left X axis is the same radial distance from the desired center but would be a rejected hole using the coordinate system. However, it would be acceptable with positional tolerancing and truly represents the hole and pin relationship at assembly. Further, if the MMC principle is applied, the holes whose produced sizes have departed from MMC size toward LMC, provide added (Bonus) tolerance; again acceptable to the design requirements. The black dot centers outside of the ∅.007 zone indicates use of such added tolerance if the hole is produced at, for example .254. The use of the datum references is also possible to ensure hole attitude (orientation) to specific surfaces as relative to part function and relationship. Further datum references for pattern location of the holes from the outer surfaces of the part would also be required to complete the requirement. See page 134 and following pages.

It is evident that the position tolerance method does provide advantages and moreover provides the only standard for universal uniform interpretation for feature location. See following pages for further discussion and added coverage of position tolerance application.

RFS APPLICATION

Where the MMC principle cannot be permitted by the design requirements, the feature control frame of the POSITION SYSTEM example would be shown as:

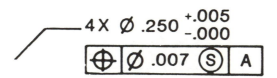

The RFS modifier (S) is stated according to RULE #2. This no longer permits the bonus tolerance. This is a more precision oriented and higher cost requirement. For example, the two holes in the black dot layout outside of the ∅.007 tolerance zone would be rejected (unacceptable) holes.

POSITION TOLERANCE THEORY

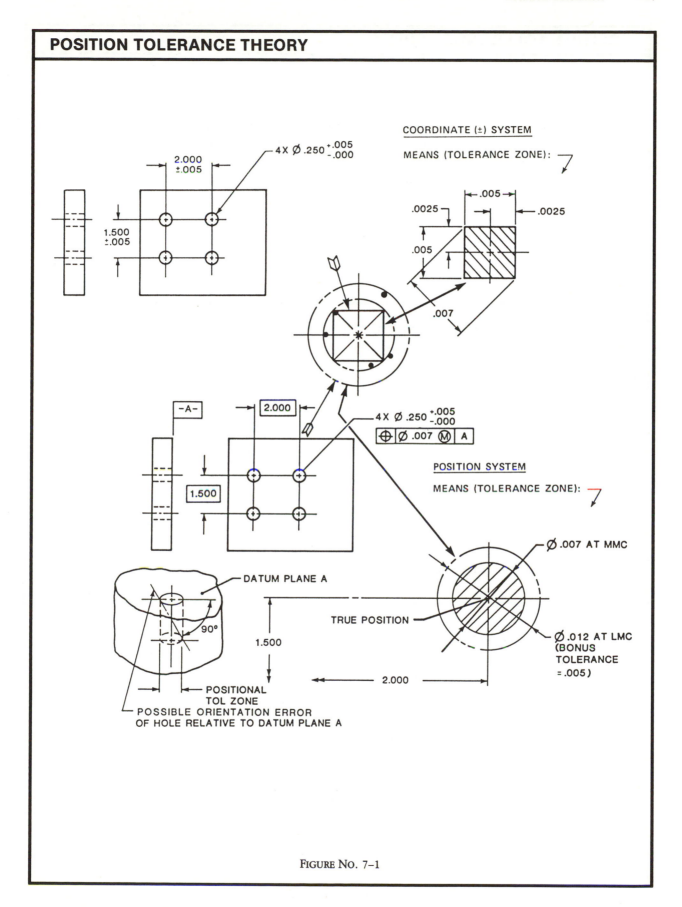

COORDINATE (±) SYSTEM

MEANS (TOLERANCE ZONE):

4X Ø .250 +.005 -.000

2.000 ±.005

1.500 ±.005

.005

.0025   .0025

.005

.007

POSITION SYSTEM

MEANS (TOLERANCE ZONE):

-A-

2.000

1.500

4X Ø .250 +.005 -.000

⊕ Ø .007 Ⓜ A

DATUM PLANE A

90°

1.500

POSITIONAL TOL ZONE

POSSIBLE ORIENTATION ERROR OF HOLE RELATIVE TO DATUM PLANE A

Ø .007 AT MMC

TRUE POSITION

Ø .012 AT LMC (BONUS TOLERANCE = .005)

2.000

FIGURE NO. 7-1

POSITION TOLERANCE—RULE 2

For a *tolerance of position*, Ⓜ, Ⓢ, or Ⓛ, must be specified on the drawing with respect to the individual tolerance, datum reference, or both, as applicable.

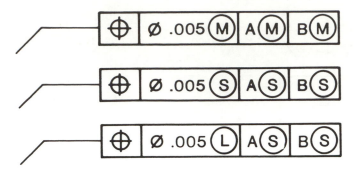

Note: Former practice (Rule #2) not recommended for new applications.*

For a tolerance of position, MMC applies with respect to an individual tolerance, datum reference, or both, where no condition is specified. RFS must be specified where it is required.

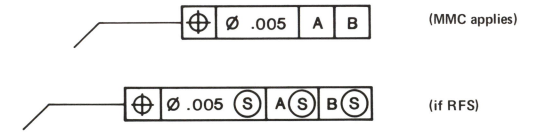

*Retained for information only to interpret drawings prepared under earlier standards (i.e., Y14.5-1973).

Note:

All examples shown in the position tolerance section and discussed in the text could, if required by the design, be applied on a regardless of feature (RFS) basis. Where such application is considered, the principles of position tolerance remain the same, except, of course, the bonus or increased tolerance possibility is no longer available. The functional gage methods are no longer valid, as well, when RFS is applied to the feature controlled. The RFS principle maintains greater precision of the concerned part features and is a valid option when the design function demands this precision and could not allow the latitude of MMC.

NOTES

DATUMS—DATUM REFERENCE FRAME (THREE PLANE CONCEPT)   │ -A- │

DEFINITION

A datum is a theoretically exact point, axis, or plane derived from the true geometric counterpart of a specified datum feature. A datum is the origin from which the location or geometric characteristics of features of a part are established.

A datum feature is an actual feature of a part which is used to establish a datum.

BASIS FOR THE DATUM REFERENCE FRAME

A datum is established from an actual part feature—a datum feature.

A datum feature refers to the actual part feature and, thus, includes all the inaccuracies and irregularities of the produced feature.

Often more than one datum feature is required on a part and a datum reference frame (also known as the three plane concept) is invoked by specified datums. Datum precedence is also to be considered where more than one datum is specified.

The datum reference frame is the application of the three mutually perpendicular (90°) planes to the part. It provides part geometry relative to the engineering design requirements and manufacturing and verification processes.

A part can be established as relative to the desired three planes by specifying the appropriate features as datum features.

A three-plane datum reference frame is typically established on a non-cylindrical part, such as shown in Fig. 7-2, by a 3-point (minimum) contact on a surface to determine the primary datum plane, 2-point (minimum) contact on a second surface to determine the secondary datum plane, and a 1-point contact on a third surface to determine the tertiary (third) datum plane. The appropriate part features are then related to these planes according to part functional requirements.

Usually, the largest and most influential surface to the part orientation and relative to the features concerned is selected as the primary datum. Part function directs which other surfaces are selected as secondary and tertiary datums.

Where parts are of irregular shape, have surface variation, or need support, datum targets may be practical in establishing the three-plane reference.

The three-planes are assumed to be mutually perpendicular (90°) to one another exactly. This is in keeping with reference to such established principles in engineering, mathematics, geometry, X-Y-Z machine and instrument movement, the three-planes used in normal inspection procedures, etc. In a practical sense, the three-planes as implemented in manufacture and inspection are envisioned as being within tooling or gaging tolerances and within human capabilities to approach exactness.

Application of the datum pick-up on parts, and in a datum sequence if required, assumes that the datum precision resides in the tool, gage, fixture, machine, surface plate, etc. It, therefore, is in the contact of the actual part with the represented simulated datum planes in such tools and equipment that establishes the datum planes and, thus, the part relationship. The processing equipment (surface plates, fixtures, etc.) used to establish the datums are referred to as simulated datums.

DATUMS—DATUM REFERENCE FRAME (THREE PLANE CONCEPT)

AS DRAWN

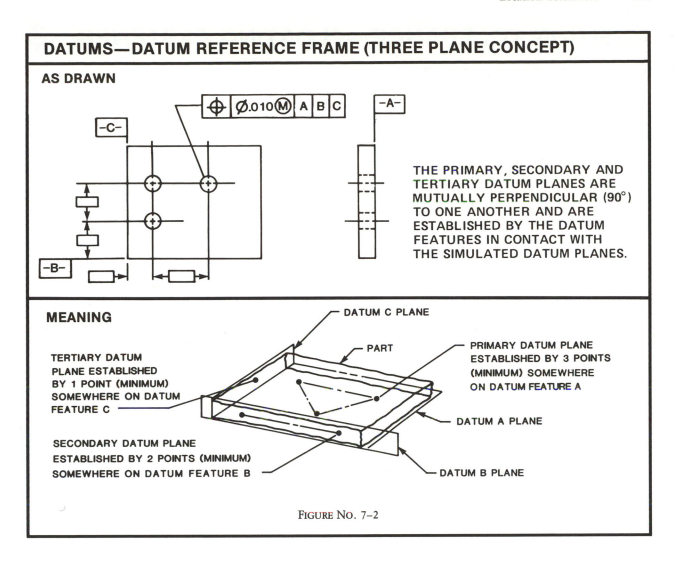

THE PRIMARY, SECONDARY AND TERTIARY DATUM PLANES ARE MUTUALLY PERPENDICULAR (90°) TO ONE ANOTHER AND ARE ESTABLISHED BY THE DATUM FEATURES IN CONTACT WITH THE SIMULATED DATUM PLANES.

MEANING

DATUM C PLANE

PART

PRIMARY DATUM PLANE ESTABLISHED BY 3 POINTS (MINIMUM) SOMEWHERE ON DATUM FEATURE A

TERTIARY DATUM PLANE ESTABLISHED BY 1 POINT (MINIMUM) SOMEWHERE ON DATUM FEATURE C

DATUM A PLANE

SECONDARY DATUM PLANE ESTABLISHED BY 2 POINTS (MINIMUM) SOMEWHERE ON DATUM FEATURE B

DATUM B PLANE

FIGURE NO. 7–2

THE DATUM FEATURE, THE DATUM PLANE, THE SIMULATED DATUM

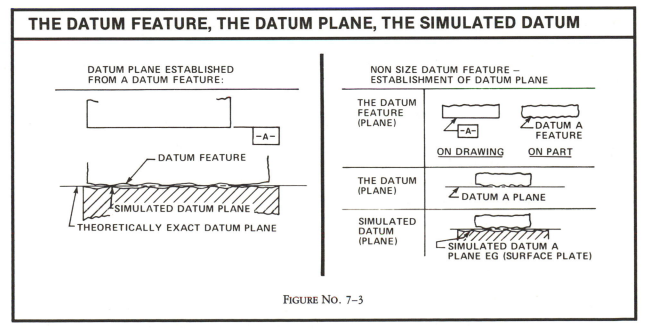

DATUM PLANE ESTABLISHED FROM A DATUM FEATURE:

-A-

DATUM FEATURE

SIMULATED DATUM PLANE

THEORETICALLY EXACT DATUM PLANE

NON SIZE DATUM FEATURE – ESTABLISHMENT OF DATUM PLANE

THE DATUM FEATURE (PLANE)	-A- ON DRAWING	DATUM A FEATURE ON PART
THE DATUM (PLANE)	DATUM A PLANE	
SIMULATED DATUM (PLANE)	SIMULATED DATUM A PLANE EG (SURFACE PLATE)	

FIGURE NO. 7–3

POSITION TOLERANCE—FLOATING FASTENER

THE DESIGN REQUIREMENT

Where two (or more) parts, both with holes, must align at the holes sufficiently to accept pins, screws, etc. in assembly, the floating fastener method and calculations may be used to determine the position tolerances for both parts.

For simplicity in explaining this principle, only one datum (the primary datum) is shown on each of the mating parts in Fig. 7-4. A complete drawing would use the 3-datum Datum Reference Frame. The corresponding features (mating surfaces), designated as datum A on both parts (top face surface of the upper part, back face surface of the lower one), are selected. Of course, they do not both need to use the datum reference letter A. Any letters are useable on either part. The material condition needed must be selected and specified according to Rule #2.

DEVELOPING AND CALCULATING THE POSITION TOLERANCE

The size of the fastener is selected (.190) and the appropriate size of the clearance holes ($\varnothing.205$ $^{+.005}_{-.000}$) are then determined and specified as based upon the designer discretion or as selected from stan- dards recommendations.

The maximum material condition sizes of both mating features, i.e., the fastener (O.D.) and the hole, are used to calculate the position tolerance for the clearance holes. The same size holes can be used on both parts to simplify calculations. If different size holes are used the calculations are repeated for each size specified.

The floating fastener calculation determines the positional tolerance for both part holes. The formula (where hole (MMC) = H, fastener (MMC) = F, and position tolerance = T) is:

$$T = H - F (\varnothing.015 = \varnothing.205 - \varnothing.190)$$

Where reverse methods may be desired as based upon a predetermined positional tolerance (such as where the hole size must be determined as based upon a given fastener size) the formula could be:

$$H = F + T (\varnothing.205 = \varnothing.190 + \varnothing.015)$$

If maximum material condition is specified, as in this example, the stated positional tolerance is increased an amount equal to the departure from MMC size as the actual hole is produced and its size is established.

Datum references are required to ensure orientation of the holes relative to the respective mating part surfaces as appropriately stated with a datum reference frame.

MEANING AND VERIFICATION PRINCIPLES

Production benefits from the clear and uniform meaning of the design requirements. Where MMC is applied, added production tolerance is possible as shown in the tabulations.

Functional gage principles may be utilized where the maximum material condition principles are specified. The gage member sizes as shown are developed from the part hole sizes. The formula is:

$$P = H(MMC) - T$$

Where the same position tolerance is used on both parts, the same functional gage may be used on both parts. Gage maker's tolerances must also be considered according to standard practices. Note that the virtual condition of the holes establishes the nominal gage pin size.

Other comparable measuring techniques may be used, such as coordinate measuring machine. If the RFS method had been specified on the drawing, the functional gage would not be possible. Coordi- nate measuring methods would then be required.

POSITION TOLERANCE—FLOATING FASTENER MATING PARTS ⊕

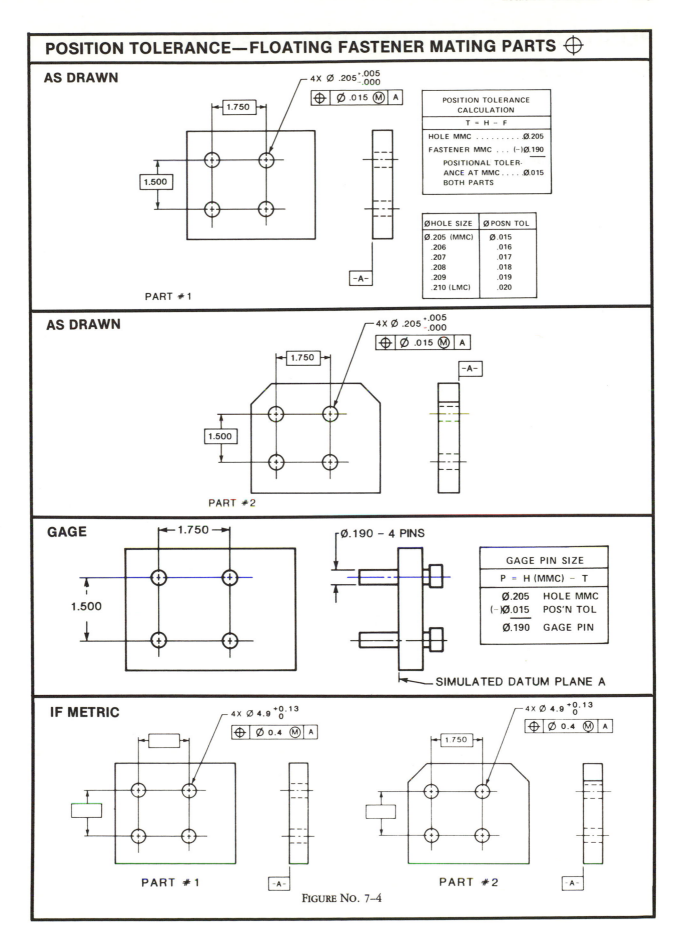

FIGURE NO. 7–4

POSITION TOLERANCE—FIXED FASTENER

THE DESIGN REQUIREMENT

Where two (or more) parts have mating features, holes and pins, but where the pin is fixed to one part but must enter the mating part hole at assembly, the fixed fastener method and calculation may be used to determine the position tolerances for both parts.

For simplicity in explaining this principle, only one datum (the primary datum) is shown on each of the mating parts in Fig. 7-5. A complete drawing would use the 3-datum Datum Reference Frame. The corresponding features (mating surfaces) designates datum A on both parts (back face of the upper part, top face of the lower one), are selected. Of course, they both do not need to use the same datum reference letter A—any letters are useable on either part. The material condition needed must be selected and specified according to Rule #2.

DEVELOPING AND CALCULATING THE POSITION TOLERANCE

The size of the fixed pin is selected. A .190-32 UNF-2A threaded pin pressed into the plate is determined as the means of assembly (Part #2). An appropriate size for the clearance holes ($\varnothing$.206) is selected as based upon the design requirements (Part #1).

The maximum material condition sizes of both mating features, i.e., the pin or fastener (O.D.) and the hole, are used to calculate the position tolerances for the pins and holes on both parts.

The fixed fastener calculation determines the positional tolerance for both parts. The formula (where hole (MMC) = H, pin (MMC) = P, and position tolerance = T) is:

$$T = \frac{H - P}{2}\left(\varnothing.008 = \frac{\varnothing.206 - \varnothing.190}{2}\right)$$

(Even though the .190-32 pin will truncate some, a safe .190 size is assumed.)

Where reverse methods may be desired, as based upon a predetermined positional tolerance (such as where the hole size must be determined as based upon a given fastener or pin size) the formula (where 2T = total of position tolerance on both parts) could be:

$$H = P + 2T \ (\varnothing.206 = \varnothing.190 + \varnothing.008 + \varnothing.008)$$

Where desirable to select a more suitable distribution of tolerance between parts, the calculated total tolerance may be divided between the parts (where .016 is the total tolerance to be distributed, such combinations as .006 and .010, .005 and .011, .007 and .009, etc.). This is done at the design stage before release to production.

If maximum material condition is specified, as in this example, the stated positional tolerances on each part are individually increased an amount equal to the departure from MMC size as the actual holes and pins are produced (this is the bonus tolerance).

Datum references are required to ensure orientation of the holes and pins and their respective mating part surfaces are appropriately stated with a datum reference frame.

Where the position tolerance in such a mating part situation is to be based upon the major diameter (O.D.) rather than the normally implied pitch diameter (i.e., per the Pitch Diameter rule), a notation such as "MAJOR $\varnothing$" must be placed beneath the concerned feature control frame. This has been done on the lower part, since the mating part hole relates to the O.D. (outside diameter) of the threaded pin, not the pitch diameter. (Refer to the Supplemental Information section, for further information on the Pitch Diameter Rule and its application (page 189).

POSITION TOLERANCE—FIXED FASTENER MATING PARTS ⌖

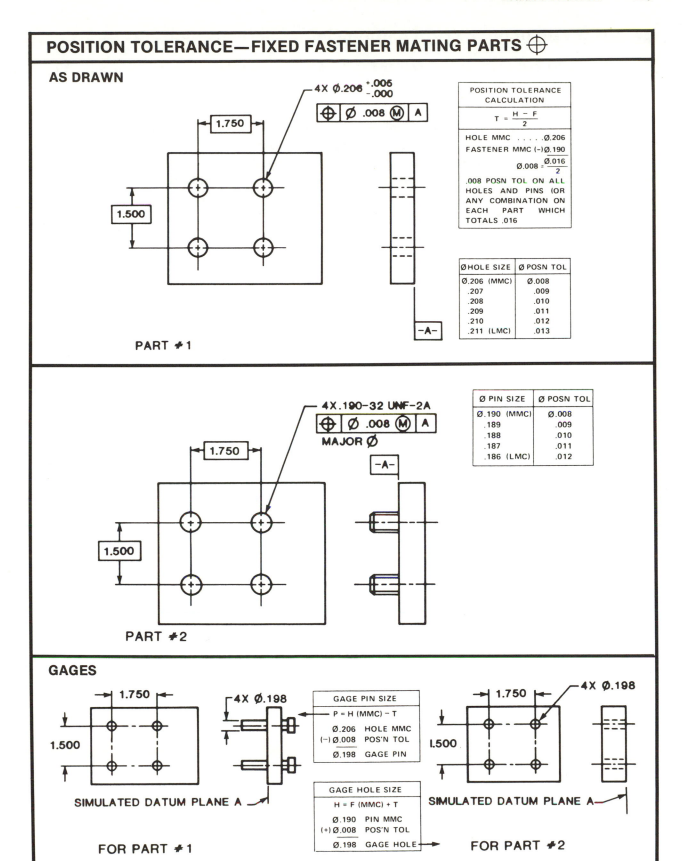

AS DRAWN

4X Ø.206 $^{+.005}_{-.000}$

| ⌖ | Ø .008 Ⓜ | A |

POSITION TOLERANCE CALCULATION

$$T = \frac{H - F}{2}$$

HOLE MMC Ø.206
FASTENER MMC (−)Ø.190

$$Ø.008 = \frac{Ø.016}{2}$$

.008 POSN TOL ON ALL HOLES AND PINS (OR ANY COMBINATION ON EACH PART WHICH TOTALS .016

Ø HOLE SIZE	Ø POSN TOL
Ø.206 (MMC)	Ø.008
.207	.009
.208	.010
.209	.011
.210	.012
.211 (LMC)	.013

1.750

1.500

PART #1

−A−

4X .190-32 UNF-2A

| ⌖ | Ø .008 Ⓜ | A |

MAJOR Ø

−A−

Ø PIN SIZE	Ø POSN TOL
Ø.190 (MMC)	Ø.008
.189	.009
.188	.010
.187	.011
.186 (LMC)	.012

1.750

1.500

PART #2

GAGES

1.750

1.500

4X Ø.198

GAGE PIN SIZE

P = H (MMC) − T

Ø.206	HOLE MMC
(−)Ø.008	POS'N TOL
Ø.198	GAGE PIN

GAGE HOLE SIZE

H = F (MMC) + T

Ø.190	PIN MMC
(+)Ø.008	POS'N TOL
Ø.198	GAGE HOLE

SIMULATED DATUM PLANE A

FOR PART #1

1.750

1.500

4X Ø.198

SIMULATED DATUM PLANE A

FOR PART #2

FIGURE NO. 7–5

MEANING AND VERIFICATION PRINCIPLES

Production benefits from the clear and uniform meaning of the design requirements. Where MMC is applicable, added (bonus) production tolerance is possible as shown in the tabulation with each part.

Functional gage principles may be utilized where the maximum material condition principles are specified. The gage member sizes as shown are developed from each part to be gaged and their respective hole and pin sizes. (Formula to determine gage pin sizes for part with holes (Part #1): P = H (MMC) – T., formula to determine gage holes sizes for part with pins (Part #2): H = P(MMC) + T.) Gage maker's tolerances must also be considered according to standard practices. It should be noted that the resulting virtual condition on each part determines the nominal gage member size.

Other comparable measuring techniques may be used, such as a coordinate measuring machine. If the RFS method had been specified in the drawing, the functional gages would not be possible. Coordinate measuring would then be required.

NOTES

POSITION TOLERANCE—FIXED FASTENER MATING PARTS—THREADED HOLES

THE DESIGN REQUIREMENT

Where two or more parts have mating features in a similar relationship as those of the preceding example (Fig. 7-5), but there is a threaded (tapped) hole in one of the parts, the fixed fastener method and calculation is again used to determine the position tolerance for both parts.

For simplicity of explaining this principle, only one datum (the primary datum) is shown on each of the mating parts (Fig. 7-6). A complete drawing would use the 3-datum Datum Reference Frame. The corresponding features (mating surfaces) are designated as datum A on both parts as selected. Of course, they both do not need to use the same datum reference letter A. Any letters can be used on either part. The material condition must be selected and specified according to Rule #2.

DEVELOPING AND CALCULATING THE POSITION TOLERANCE

The size of the fastener is selected—in this case a .190-32 UNF 2A screw. Thus, the threaded holes in Part #2 are determined as four .190-32 UNF-2B holes. An appropriate size for the clearance holes in Part #1 is selected at $\varnothing.206^{+.005}_{-.000}$.

The maximum material condition sizes of both mating features, i.e. the fastener (O.D.) and the hole, are used to calculate the position tolerances for both the threaded holes and the clearance holes on both parts. Since the .190-32 fastener will ultimately become fixed in assembly with the .190-32 threaded hole, we use the fixed fastener method. The fastener will seek location and contact at the pitch cylinder (pitch diameter) and thus become fixed in its location when seated. However, the mating part clearance hole must clear the fastener outside diameter (O.D.) as seen in the assembly illustration below.

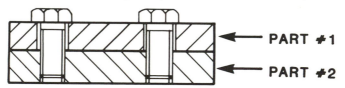

PART #1

PART #2

The results of the fixed fastener calculation determine the positional tolerance for both parts. The formula (where hole (MMC) = H, fastener O.D.(MMC) = F, and position tolerance = T) is:

$$T = \frac{H - P}{2} \left(\varnothing.008 = \frac{\varnothing.206 - \varnothing.190}{2} \right)$$

Even though the .190-32 fastener will be truncated some (slightly smaller), a safe .190 size is assumed.

Where reverse methods may be desired, as based upon a predetermined positional tolerance (such as where the hole size must be determined as based upon a given fastener size) the formula (where 2T = total position tolerance on both parts) could be:

$$H - F = 2T \ (\varnothing.206 = \varnothing.190 = \varnothing.008 = \varnothing.008)$$

Where desirable to select a more suitable distribution of tolerance between the parts, the calculated total tolerance may be divided between the parts. (where 0.16 is the total tolerance to be distributed, such combinations as .006 and 0.10, .005 and .011, .007 and .009, etc). This is done at the design stage before release to production.

If maximum material condition is specified, as in this example, the positional tolerances are individually increased. The amount of departure from MMC on the Part #1 clearance holes would be the amount of

POSITION TOLERANCE—FIXED FASTENER MATING PARTS THREADED HOLES

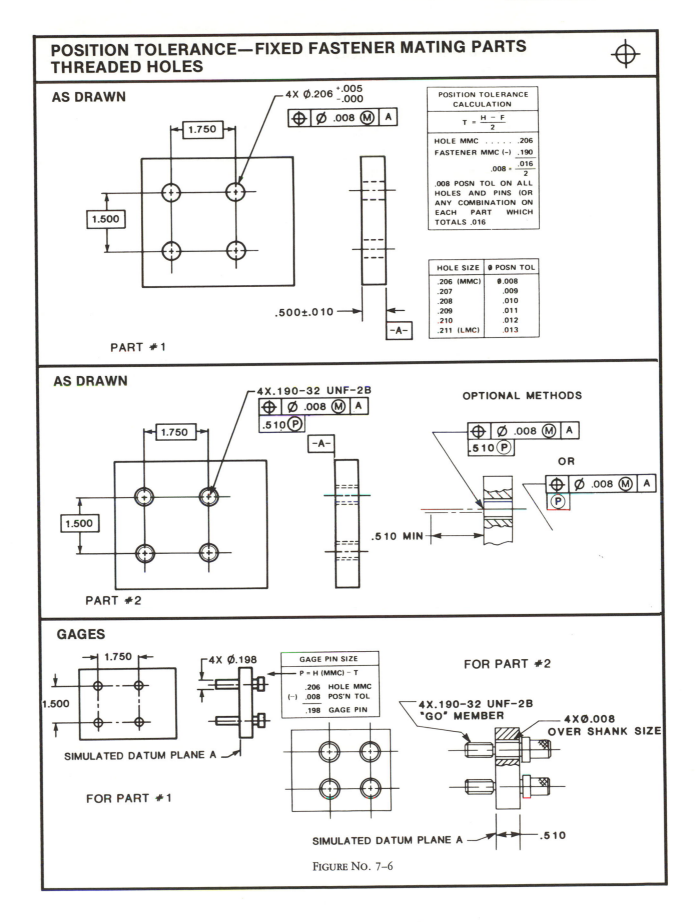

AS DRAWN

4X Ø.206 +.005 −.000

⊕ | Ø .008 Ⓜ | A

1.750

1.500

.500±.010

PART #1

POSITION TOLERANCE CALCULATION		
$T = \dfrac{H - F}{2}$		
HOLE MMC		.206
FASTENER MMC (−)		.190
.008 = $\dfrac{.016}{2}$		.016

.008 POSN TOL ON ALL HOLES AND PINS (OR ANY COMBINATION ON EACH PART WHICH TOTALS .016

HOLE SIZE	Ø POSN TOL
.206 (MMC)	Ø.008
.207	.009
.208	.010
.209	.011
.210	.012
.211 (LMC)	.013

−A−

AS DRAWN

4X.190-32 UNF-2B

⊕ | Ø .008 Ⓜ | A
.510 Ⓟ

1.750

1.500

−A−

OPTIONAL METHODS

⊕ | Ø .008 Ⓜ | A
.510 Ⓟ

OR

⊕ | Ø .008 Ⓜ | A
Ⓟ

.510 MIN

PART #2

GAGES

1.750

1.500

4X Ø.198

GAGE PIN SIZE	
P = H (MMC) − T	
.206	HOLE MMC
(−) .008	POS'N TOL
.198	GAGE PIN

SIMULATED DATUM PLANE A

FOR PART #1

FOR PART #2

4X.190-32 UNF-2B "GO" MEMBER

4XØ.008 OVER SHANK SIZE

SIMULATED DATUM PLANE A   .510

FIGURE NO. 7–6

additional (bonus) tolerance as the actual holes are produced. The amount of additional (bonus) tolerance on the threaded holes is not predictable. However, it would be some amount. Because of the centering effect of the flank angle of the screw thread form, some added tolerance is negated due to the fastener sliding towards center as it bottoms and tightens in its fastened assembly.

Datum references are required to ensure orientation control of the holes and pins and their respective mating part surfaces as appropriately stated with a datum reference frame.

In this case, the Pitch Diameter Rule is invoked automatically as applicable to the position tolerance requirement. Because the fastener will seek location at the pitch circle (pitch diameter), the basis for the requirement must be the implied pitch circle. See pages 132, 133 for continued discussion on the projected tolerance zone

MEANING AND VERIFICATION PRINCIPLES

Production benefits from the clear and uniform meaning of the design requirements. Where MMC is applicable, added (Bonus) production tolerance is possible as shown in the tabulations on Part #1. The possible added (bonus) tolerance on Part #2 can be achieved through the use of a functional gage (as shown) or comparable techniques.

The functional gage for Part #1 would be as shown in Fig. 7-6 (bottom) and as explained previously. The functional gage for Part #2 would be as shown in Fig. 7-6 (bottom) and in the following section.

NOTES

⊕ POSITION TOLERANCE—PROJECTED TOLERANCE ZONE Ⓟ

DESIGN CONSIDERATIONS

Projected tolerance zone is a method of applying position or perpendicularity tolerances which represents the mating interface of mating parts at assembly. Verification of the concerned features to its stated tolerance will predict whether the parts will assemble satisfactorily or not.

The projected tolerance zone "projects" the normal position or perpendicularity tolerance zone above (or below) the part as indicated by the call-out or other necessary detail.

The fixed fastener method of calculation is used to determine the positional tolerance when applying projected tolerance zone methods. Adding the projected tolerance zone specification moves the tolerance zone to the indicated direction and extent.

Where the projected tolerance specification requires clarification as to direction of projection on a thru hole, unique extent of projection, etc., a detail view illustrating such requirements is shown. Conventional projected tolerance zone application, where intent is clear, requires only the symbolic call-out.

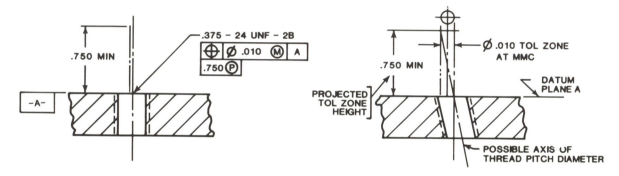

THE PROJECTED TOLERANCE ZONE

As can be seen from the cross-sectional view through the assembly of Parts #1 and #2 on page 126, perpendicularity of the threaded hole, Part #2, and thus the inserted fastener, is very critical to the assembly with Part #1. Since the position tolerance zone is 3-dimensional, an extreme case could permit the axis of the threaded hole to be out-of-perpendicular the entire ⌀.008 (at MMC) and yet be within tolerance. With even less than an extreme case, if both Parts#1 and #2, result in an out-of-perpendicularity situation in opposite directions, an interference or undue stressing of the fastener and parts could possibly occur (see next page). This situation depends of course, on numerous unpredictable results of manufacture.

POSSIBLE INTERFERENCE OR STRESSING OF PARTS AND FASTENER

By specifying the Projected Tolerance Zone on Part #2 as shown In Fig. 7-6, the ⌀.008 tolerance zone is projected above the part perpendicular to datum plane A and into Part #1. This eliminates the potential problem as shown.

The reasoning for the use of the projected tolerance zone is evident on the mating parts on page 126. Technically, the projected tolerance zone should be used in every application where a pin or fastener is to ultimately assemble in a fixed manner and project into another part or have some similar function. However, it is used on a selective basis since the interference possibility in many cases is negligible and complication of the requirement is not needed. Therefore, the specifying of the projected tolerance zone is considered optional. A rule-of-thumb could be to consider its use where the depth of the threaded hole is one diameter (of the hole) or more in depth. The amount of the projected tolerance zone is also to be chosen to best suit the design requirements. In Fig. 7-6, the maximum thickness of the mating part (Part #1) at .510 was selected.

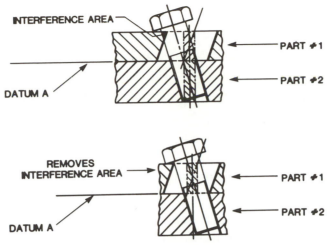

TO SHOW THE PROJECTED TOLERANCE ZONE

Where a ''through hole'' is involved (which side is projection from?) or some unique suspension above the part of the tolerance zone is required, a drawing view may be selected to show the tolerance zone placement. A thick chain line is shown offset from the hole axis (center line) with the extent of the zone given (usually as a MIN, minimum value). See the optional method as shown on Part #2 and the example on page 129.

THE FUNCTIONAL GAGE

The functional gage for Part #2, as shown In Fig. 7-6 utilizes the projected tolerance zone in constructing the gage. Note how the gage simulates Part #1. Threaded gage pins with a shank size Ø .008 undersize to the selected hole size for the gage plate are inserted and threaded into the produced Part #2 while stabilized against the datum A feature (thus the simulated datum plane).

OPTIONS TO PROJECTED TOLERANCE ZONE

Other methods which could accomplish somewhat near a comparable result to projected tolerance zone are:

   a. Calculate the amount that the conventional position tolerance zone needs to be reduced in order to compensate for the possible interference.

   b. Enlarge clearance between pin and hole and thus the resulting tolerance.

   c. Add a perpendicularity tolerance to refine the position tolerance.

POSITION TOLERANCE ⊕ —SPECIFIED DATUMS

DATUMS MUST BE SPECIFIED WITH POSITION TOLERANCE

Where a position tolerance is applied to a feature, or features (i.e., holes, pins), and their relationship is to outside edges (or other features), such features are specified as datums. This will ensure that the functional design requirements are clearly stated and uniformity of understanding in manufacturing and verification will follow. Figure 7-7 illustrates application of specific datum features.

Specifying datums also invokes datum precedence and direct relationships between the features involved.

Reading left-to-right in the feature control frame the datums specified are the primary, secondary, and tertiary (third) datums as relative to part function. Such relationship builds upon and utilizes the datum reference frame concept inherent with geometrics. Manufacturing and verification, tooling, gaging, and processing can thus be applied in keeping with the stated design requirements.

Specifying datum features may require spending more time on the design. But the benefits in clarity and production follow-through provide the best and most economic communication from design to manufacture. See also Supplemental Information section page 178.

IF METRIC APPLICATION

A comparable metric application to the subject part in Fig. 7-7.

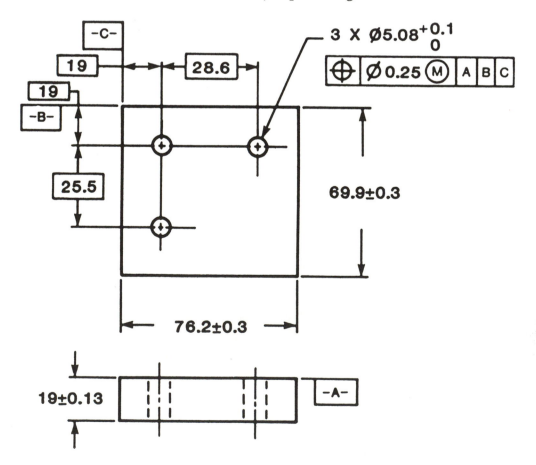

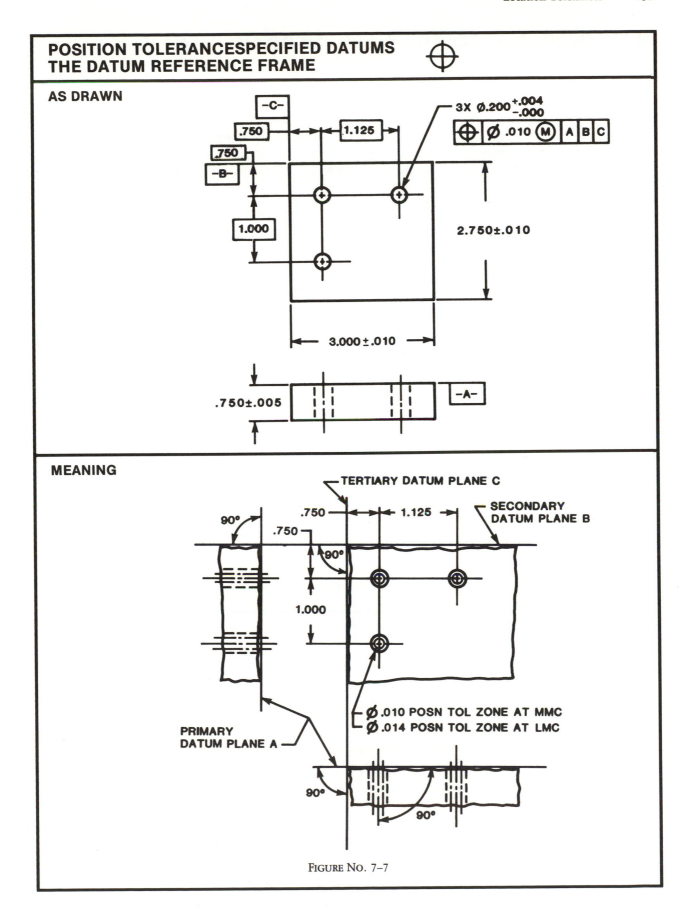

POSITION TOLERANCE SPECIFIED DATUMS THE DATUM REFERENCE FRAME

AS DRAWN

-C-

.750   1.125

.750

-B-

1.000

2.750±.010

3.000±.010

3X Ø.200 +.004 −.000

⊕ | Ø .010 Ⓜ | A | B | C

.750±.005

-A-

MEANING

TERTIARY DATUM PLANE C

SECONDARY DATUM PLANE B

90°   .750   1.125

.750

90°

1.000

90°

Ø .010 POSN TOL ZONE AT MMC
Ø .014 POSN TOL ZONE AT LMC

PRIMARY DATUM PLANE A

90°

90°

FIGURE NO. 7–7

COMPOSITE POSITION TOLERANCE

THE DESIGN REQUIREMENT

Where a position tolerance is to control the precision of features in a pattern (i.e., holes) but the pattern relationship to other features is less critical, the composite position tolerance method may be used. In such applications, the required precision in the pattern (feature relating tolerance) can be stated, yet the pattern as a unit may be separately controlled with a more lenient control relative to the part edges (datum features) with a pattern locating tolerance. The MMC principle is usually most appropriate in such applications.

Fig. 7-8 illustrates an application of composite position tolerance using the MMC principle.

The features (holes) may individually vary from true position within the specified feature relating position tolerance (Ø.008) and the established tolerance zones at each true position. The hole pattern relative to the specified datums (A, B and C) may shift/rotate from true position within the specified pattern locating position tolerance (Ø.030). These two requirements are both applicable to the feature pattern but are separate requirements.

PLACING THE REQUIREMENT ON THE DRAWING

The feature control frame is constructed as a composite symbol with the pattern locating position tolerance in the upper portion and the feature relating position tolerance in the lower portion of the symbol. There is no significance as to whether each portion is in the upper or lower segment. The datums and the tolerance values are the key indicators.

It should be noted that in the composite positional tolerancing method, the same type of control (i.e., position) is used on both portions of the requirement. Only one geometric characteristic symbol is used.

Where the composite positional tolerancing method is used, datums are required. The datum reference frame and datum precedence is also used. This ensures proper functional interface with the component or part which mounts to the located features and on the indicated datums.

CALCULATING COMPOSITE POSITION TOLERANCES

Suppose that in Fig. 7-8 the pattern locating tolerance, Ø.030, is selected as the tolerance to position the 4-hole pattern in the lower left corner of the part with some, but not critical, accuracy. The feature relating tolerance is calculated as based upon the mating part functional relationship using either the floating fastener or fixed fastener methods described previously.

MEANING AND VERIFICATION PRINCIPLES

A clear understanding of the drawing requirements and the MMC principle application will assist production.

Functional gages, as shown, may be used when composite positional tolerance is applied on an MMC basis. Two separate gages would normally be used. The gage for the pattern-locating positional tolerance (Ø.030 fig. a) would include pick-up of the datum surfaces in appropriate manner and with the virtual condition and nominal gage member size determined by MMC size of the feature (hole) minus the stated positional tolerance. The formula (where GP = Gage Pin, H = Hole, and T = Position Tolerance) is:

$$GP = H - T \ (\varnothing.176 = \varnothing.206 - \varnothing.030)$$

POSITION TOLERANCE—COMPOSITE ⌖

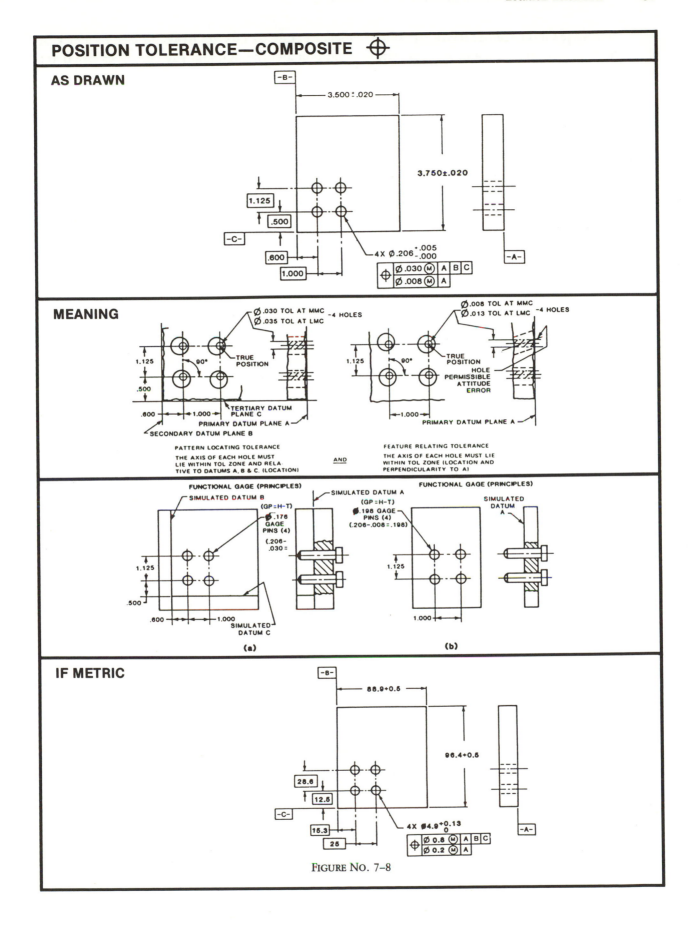

FIGURE NO. 7–8

The gage for the feature relating positional tolerance (i.e., ∅.008 fig. b) would include a pick-up of the primary datum (only) and with the virtual condition and the nominal gage member size determined by MMC size of the feature (hole) minus the stated positional tolerance. The formula is:

$$GP = H - T \, (\varnothing.198 = .206 - \varnothing.008)$$

Other comparable measuring techniques may be used, such as a coordinate measuring machine. If the RFS method had been specified in the drawing, the functional gages would not be possible. Coordinate measuring would then be required.

—————— **NOTES** ——————

POSITION TOLERANCE ⊕
COORDINATE MEASURING—GRAPHIC ANALYSIS

Graphic analysis or paper gaging is a technique which can be used to verify MMC positional tolerances using conventional measuring equipment such as a coordinate measuring machine. It can effectitely translate such measuring results into position tolerance geometry for analysis.

The part is set-up appropriately to the datum references and each feature (hole) is measured in X and Y coordinates using conventional methods (RFS).

The derived X and Y readings are compared to the desired locations as indicated on the drawing. The differential results are plotted on a piece of graph paper using an appropriate scale (e.g. one square = .001) and relative to an established true position point on the chart (can be marked 0). Each plot is made with reference to the same composite true position. See Figure 7-9 (upper illustration).

An overlay comprised of to-scale (same as graph paper) concentric circles on tracing paper, or stable plastic material, is superimposed over the plotted results of the measuring process. If the features (holes) are located from the edge datums by basic dimensions, the center of the concentric circles overlay is placed on the true position (marked 0). If the resulting plotted centers (axes) then fall within the stated drawing positional tolerance zone, such hole centers have met the requirement. In this case, there is no need to apply MMC principles in the inspection process.

Where the resultant plotted center is found outside the drawing stated positional tolerance circle, the size of the feature (hole) is determined. Its size departure from MMC enlarges the permissible positional tolerance zone. If the feature (hole) plotted center is then within the appropriate larger concentric circle (tolerance zone), the feature position is acceptable. The MMC principle is invoked only when necessary with feature centers (axes) measured RFS.

Where the features (holes) in the pattern are to be independently verified relative to their true positions, one feature (hole) may be selected as an origin of the measurements. A second feature is selected for square-up or orientation of the pattern with the balance of the features measured from the set-up using standard coordinate measuring techniques. The differential results of the X and Y measurements are then compared to the desired locations and plotted on a piece of graph paper. See Figure 7-9 (lower illustration). An appropriate scale is used (e.g. 4 squares = .001 in this example).

An overlay comprised of to-scale (same as graph paper) concentric circles on tracing paper or stable plastic material, is superimposed over the plotted results of the measuring process. If the overlay circle equivalent of the stated position tolerance can encompass all the plotted measurement results simultaneously, the features' (holes') locations have met the requirement. Where necessary, the size of an individual feature (hole) must be determined to enlarge that permissible tolerance zone (larger equivalent overlay circle) to re-evaluate its acceptability. If all plotted locations can be contained with their allotted tolerance zones simultaneously in any one position of the overlay, the features are acceptable as individual features and a pattern. Note that the overlay center of the concentric circles is *not* placed on the true position (marked O). It is placed at any appropriate place as above described.

Paper gaging provides a method of simulating part assembly and duplicates the principles of functional gaging with open set-up techniques. Paper gaging can often be used to manipulate coordinate measuring results, such as hole-to-hole (lower illustration in Fig. 7-9). If a computer program is not available, such methods can effectively perform a rather sophisticated analysis (hole-to-hole).

Programmable calculators and computer programs with hole-to-hole analysis capability could also perform such a verification procedure while by-passing a functional gage or graphic analysis.

The Conversion Table in the Supplemental section (Page 193) can also be used to determine acceptance of parts from coordinate measuring results on Figure 7-9 (upper illustration). The Table *cannot* be used on Figure 7-9 (lower illustration) analysis.

POSITION TOLERANCE—COORDINATE MEASURING— GRAPHIC ANALYSIS

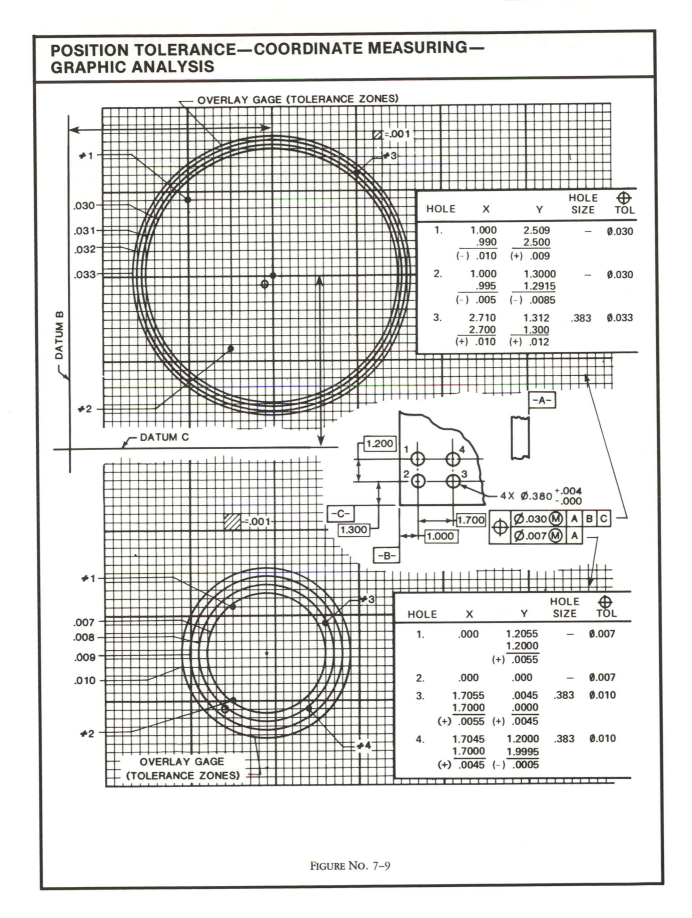

HOLE	X	Y	HOLE SIZE	⊕ TOL
1.	1.000	2.509	—	Ø.030
	.990	2.500		
	(−) .010	(+) .009		
2.	1.000	1.3000	—	Ø.030
	.995	1.2915		
	(−) .005	(−) .0085		
3.	2.710	1.312	.383	Ø.033
	2.700	1.300		
	(+) .010	(+) .012		

HOLE	X	Y	HOLE SIZE	⊕ TOL
1.	.000	1.2055	—	Ø.007
		1.2000		
	(+)	.0055		
2.	.000	.000	—	Ø.007
3.	1.7055	.0045	.383	Ø.010
	1.7000	.0000		
	(+) .0055	(+) .0045		
4.	1.7045	1.2000	.383	Ø.010
	1.7000	1.9995		
	(+) .0045	(−) .0005		

FIGURE NO. 7–9

POSITION TOLERANCE ⊕ —SIZE FEATURE DATUM

THE DESIGN REQUIREMENT

Where position tolerance is applied to features in a pattern (i.e., holes) and the pattern relationship is to another feature, such as a pilot hole, that feature can be indicated as a locating datum. In such a case, the location of the surrounding feature pattern relative to the pilot hole is the critical requirement. A mating part situation with a pilot pin surrounded by its counterpart features (pins, tapped holes) can be envisioned as the mating part interface. See Figure 7-10.

The pilot hole feature may first be specified with a more lenient positional tolerance relative to the selected outside features. Where necessary, a refinement in orientation (i.e., a perpendicularity tolerance) may be necessary to ensure the proper pilot hole orientation and pilot pin mating part interface. Since the datum feature is a size feature, the MMC principle can be applied if appropriate to the design requirement; for example, if there is to be a clearance fit between the pilot hole and pilot pin assembly.

Where the surrounding holes are to interface with mating part features their positional tolerance is calculated using the fixed fastener formula and maximum material condition is specified. The location dimensions for the surrounding holes are specified relative to the pilot hole.

The datum references specified with the surrounding holes are, first, the orientation datum (top surface) as the primary datum, the location datum (pilot hole) as the secondary datum, and an outside surface as the tertiary datum.

MEANING AND VERIFICATION PRINCIPLES

As indicated by the Datum/Virtual Condition Rule, the pilot hole is implied at its virtual condition. That is, the pilot hole has been permitted orientation tolerance in its control. This, therefore, must be recognized in its "pick-up" in fixturing and inspection and as pertinent to the part function.

Where the maximum material condition is specified to the surrounding holes and also to the pilot hole, the full advantages of MMC are realized. Part function is assured, additional production tolerance is available, and functional gaging techniques may be used. As each surrounding hole size departs from its MMC in production, an increase in the hole position tolerance is realized to the extent of that departure. As the pilot hole size departs from its MMC in production, a shift of the surrounding hole pattern as a group is permissible relative to the pilot hole.

Functional gaging is permissible when MMC is specified. Functional gaging would simulate the mating part interface, expedite inspection operations, and effectively capture the interplay between feature size and location. Feature sizes must be verified separately and independently. Open set-up measuring techniques can, of course, be used in lieu of functional gaging with uniform results.

Where RFS principles might be invoked, such as the surrounding hole relationship to the pilot hole at RFS, any variation in pilot hole size will not permit a shift in the pattern relative to the pilot hole. That is, the design requirement prohibits such a shift. Any functional gaging or inspection processes must also carry out such restrictions.

GRAPHIC ANALYSIS AND COMPUTER ANALYSIS

Graphic analysis (paper gaging) can be used to determine acceptance of parts as shown. In such an application, however, the true position tolerance zones are about the datum axis as a center with float of the 4 hole pattern (as a group) permitted in a zone equal to the departure from MMC of the datum feature. Special computer software programs can perform such analysis thus eliminating the need to paper gage or functionally gage the part.

POSITION TOLERANCE—SIZE FEATURE DATUM ⌖

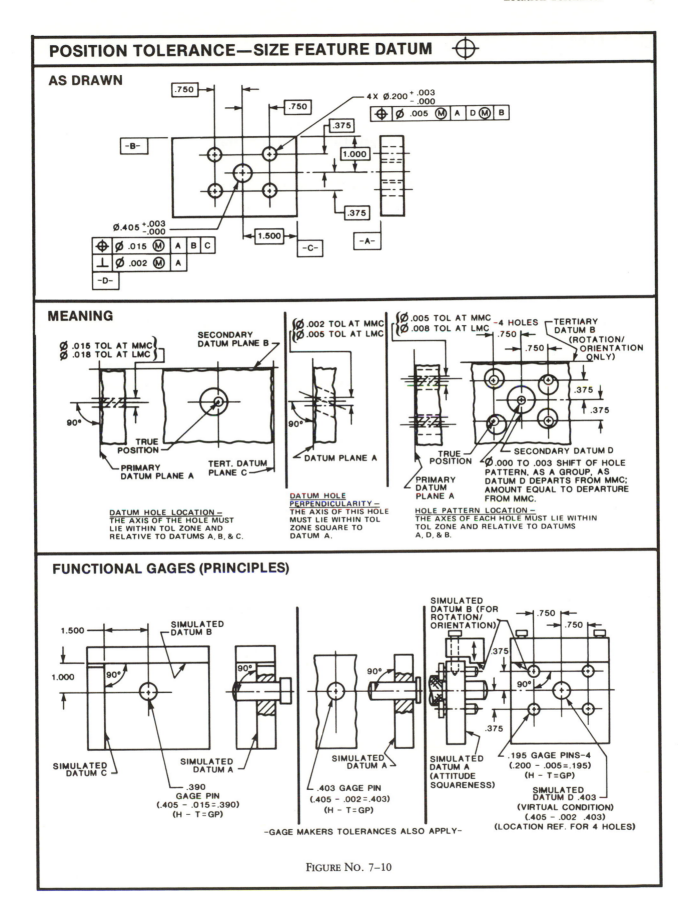

FIGURE NO. 7–10

DATUM TARGETS   ⌬

DESIGN AND PRODUCTION CONSIDERATIONS

It may be necessary to use datum targets to establish datum planes on parts made from castings, forgings, sheet metal, weldments, etc. where surfaces or features are of less dependable precision. The datum targets are used to establish the necessary datum reference frame for geometric tolerancing on such parts.

Datum targets are designated by a circle divided into halves with the datum letter (of datum plane, etc. being established) at lower left, and number of the target at lower right in the circle. Numbering of the symbol should be sequential in each plane being established.

Datum targets may be of three types: points, lines, or areas on the part. They are designated by datum target symbols leading to the part surface and terminating at a point or line (marked by X) or an area of shape and size indicated. The upper half of the symbol may be used to indicate the datum target area size. See Fig. 7-11, upper illustration.

Datum targets are normally not features but only points, lines, or areas of pick-up or location on the part features or surfaces.

Datum target locations and size (if an area) are usually designated as ''basic'' $\boxed{.XXX}$, implying standard tooling, gaging, set-up or size tolerances when implemented into the manufacturing tools, gages, or procedures. If desired, datum targets may instead be given specified location (and/or size) tolerances although caution should be exercised since the target can then be possibly misunderstood as a part feature requirement.

Datum targets tend to imply manufacturing information is being placed on the engineering drawing and it can be construed as such in many cases. However, its value in ensuring consistency between key manufacturing and verification processes and, thus, ultimately protecting total design integrity, makes datum targeting an acceptable drawing convention. Also, datum target methods often are most effective and necessary to state unique design requirements. Thus, the technique is also a viable design tool.

Datum targets ''construct'' the necessary datum planes. Therefore, to ensure that sense of construction and clarify intent, the conventional datum feature symbols are also used to show the resulting planes.

Implementation of targeting in manufacture, fixturing, etc. are represented by appropriate spherical radius pins, tangent contact of round pins, or flat ended pins as indicated by the target itself. See Fig. 7-11, lower illustration, for an example of target application and lower right for tooling follow-through.

DATUM TARGETS

AS DRAWN

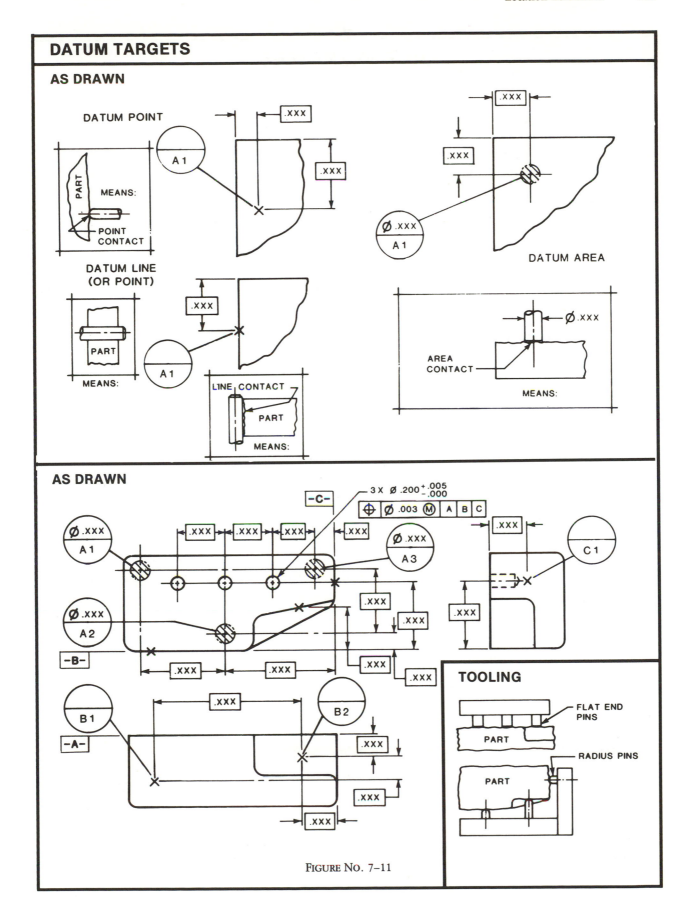

FIGURE NO. 7–11

7
QUIZ-EXERCISES
LOCATION/POSITION TOLERANCES

The following questions are relative to the material in this chapter. Read the question and answer to the best of your ability. The answers can be found in the companion manual *Answer Book and Instructor's Guide for Introduction to Geometric Dimensioning and Tolerancing*.

Consult your instructor if you have any questions.

1. Tolerances of location involve the use of geometric characteristics _____ and _____. Which characteristic is used only on an RFS basis? _____.

2. Location tolerances are applied to (size/non-size) _____ features and control _____ between two or more features.

3. Location tolerances relate their tolerance zones to which of the below?

   _____ surface

   _____ axis

   _____ centerplane

   _____ centerline

4. Where function or interchangeability of mating part features is involved, the (RFS/MMC) _____ principle and _____ tolerancing may be introduced to great advantage.

5. Position tolerancing is a method used to specify the location of features with respect to one another and in relationship to a _____ feature.

6. A position tolerance is the total permissible variation in the location of a feature about its desired or exact _____.

7. For cylindrical features (holes and bosses), the positional tolerance is a _____ _____ shaped tolerance zone within which the _____ of the feature must lie. The shape of the tolerance is specified with the symbol _____.

8. Position tolerance is a (cumulative/non-cumulative) _____ method of control in which each feature relates to its own desired exact (true) position.

9. Position tolerances are normally based on the _____ size of the concerned feature as it relates to the _____ of the corresponding mating part feature.

10. When position tolerance on an MMC basis is applied to a feature, the tolerance on the actually produced feature (increases/decreases) _____ as the size departs from MMC size.

11. Position tolerance is ideally suited to multiple mating part cylindrical features in a pattern. Based upon part function, which material condition is best to ensure interchangeability and with maximum tolerance advantage? _____. Which material condition would be used for higher precision? _____.

12. Which two of the three statements below most support the substance of the preceding questions and are the best technical reasons for using position tolerancing? (Place a check at your choices).

_____ Position tolerancing recognizes the permissible variation of a cylindrical feature location in 360° of movement.

_____ Position tolerancing is a more convenient way to relate mating features than plus and minus coordinate tolerancing.

_____ The position tolerance is developed directly from the relationship of the mating feature MMC sizes.

13. In this drawing, what is the MMC size of the Ø.250 holes? _____. Specify on the drawing that the four Ø.250 holes are to be located within a position tolerance of .010 diameter at MMC.

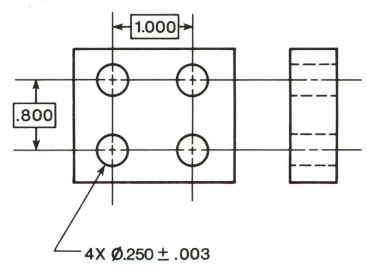

4X Ø.250 ± .003

14. On the figure shown under question 13, what is the position tolerance of the hole if it is produced at Ø.247? _____, at Ø.253? _____.

15. On this layout of the part under question 13, sketch in the position tolerance zones at MMC size of the hole and at LMC limit size of the hole.

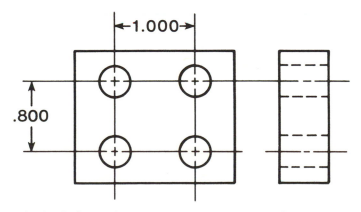

16. Datum references as the basis for position relationships are (specified or implied) _____ on the drawing. Datum selection should be relative to part function and those features selected, called _____ features, should be _____ features.

17. On the below part, suppose that the surface shown in the front view is the mounting face and the other two surfaces are important to the hole pattern positional tolerance. Add three datums with precedence indicated.

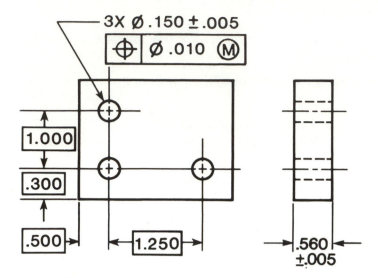

18. Make a sketch below showing the interpretation of the part under the preceding question as based upon your answer.

19. On the part below add a specification on the Ø.376 hole to show it located at true position within .010 diameter at MMC with respect to datums A, B and C. Indicate datum precedence. Also add a perpendicularity requirement (within Ø.003 at MMC) as a refinement of the position tolerance. Use .750 (vertical) and .940 (horizontal) basic dimensions.

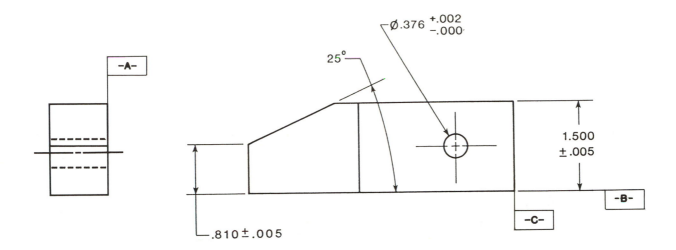

20. Assume that the two parts below are mating parts with the four holes in each to coincide so that four .138-32 screws will assemble. Calculate the position tolerances and complete the position dimensioning and tolerancing on the two parts. Disregard hole pattern location from the edges but determine a proper orientation (primary) datum on each part.

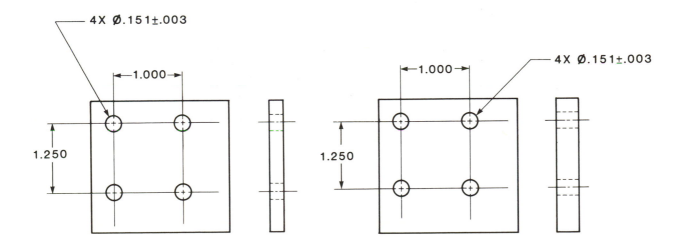

21. The parts shown are mating parts to be assembled with two .138-32 screws and located upon the two dowel bosses and mating holes. Calculate the position tolerances and complete the position dimensioning and tolerancing, including establishing three datums on each part.

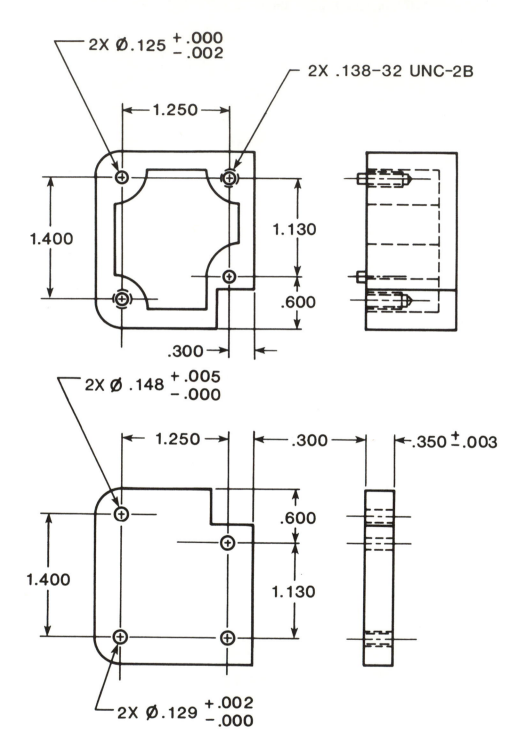

22. Unless otherwise specified, the position tolerance zone extends to the _____ of the feature.

23. In the example in question 21, could the distribution of the position tolerance on each part have been different? _____. To support your answer, which of the below statements is most appropriate?

    _____ Position tolerance is calculated on the basis of relationship of size of the corresponding mating part features.

    _____ The total position tolerance calculated may be distributed as desired between the corresponding mating part features.

24. Referring to the "answer" illustration in question 21, what is the position tolerance of the Ø.125 boss if produced at Ø.1235? _____. The Ø.148 hole if produced at Ø.151? _____. Why is the position tolerance of the .138-32 hole different in this regard? Select most appropriate answer:

    _____ Tapped holes usually have close size tolerances.

    _____ The centering effect of the inserted screw may negate added tolerance due to size deviation from MMC.

25. Make an analysis (paper gage) of the part shown on the following page as based upon the given inspection results from a coordinate measuring operation on a produced part. The results are X and Y measurements of the produced holes from the datum planes and the actual hole sizes. Make the necessary calculations and plot (use dots) the results on the graph using the zero (0) point as the true position and origin for the X and Y differentials. Imagine the concentric circles shown on the graph as a transparent overlay chart (the tolerance zones) of the same scale as the graph and placed at true position after plotting the hole centers. The graph scale is 1 square = .001 inch. Number the holes #1 (upper left), #2 (lower left), #3 (upper right), and #4 (lower right).

	Measurement X Direction	Measurement Y Direction	Hole Size
Hole #1	.749	2.2525	Ø.3025
Hole #2	.752	.744	Ø.303
Hole #3	2.746	2.2457	Ø.302
Hole #4	2.7454	.7552	Ø.304

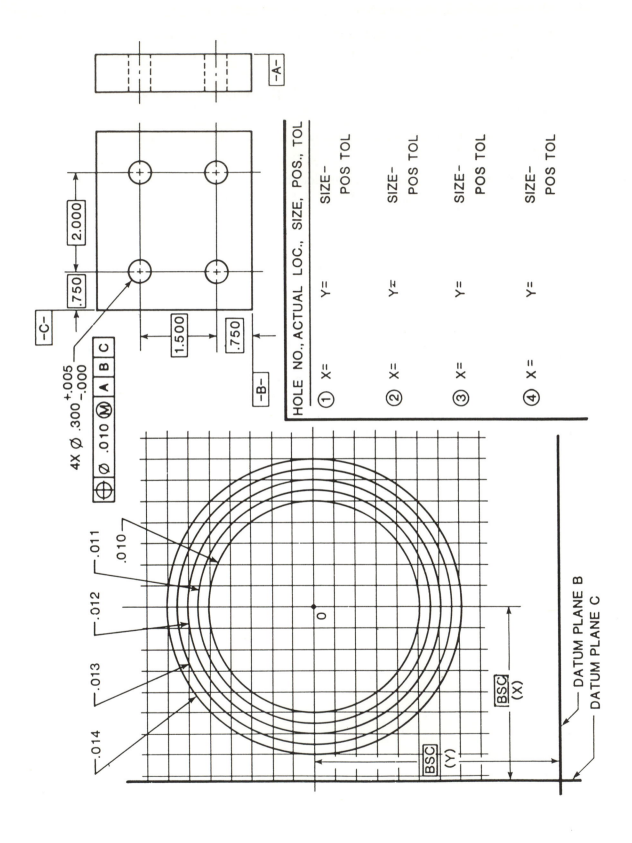

HOLE NO., ACTUAL LOC., SIZE, POS., TOL		
① X= Y=	SIZE – POS TOL	
② X= Y=	SIZE – POS TOL	
③ X= Y=	SIZE – POS TOL	
④ X = Y=	SIZE – POS TOL	

26. From your analysis of the part under question 25, is the part acceptable? Yes or No? _____

27. Using your calculator*, confirm your answer to question 25 mathematically. Fill in the derived diametrical (cylindrical) values calculated for each hole (show at least to the fifth decimal place).

    Hole #1 _____

    Hole #2 _____

    Hole #3 _____

    Hole #4 _____

    (*If no calculator is available to you, do the best you can with the tables, graphs, and calculation methods shown in your reference materials.)

28. Using only your calculator*, determine from the inspection results of another part (produced to the same drawing as show under question 25) if the part is acceptable.

	Measurement X Direction	Measurement Y Direction	Hole Size
Hole #1	.753	2.248	Ø.302
Hole #2	.745	.752	Ø.301
Hole #3	2.7545	2.254	Ø.303
Hole #4	2.7445	.746	Ø.303

    Fill in the derived diametrical (cylindrical) values calculated for each hole (show at least to the fifth decimal place). Indicate at right acceptance or rejection of each hole. (yes or no)

    Hole #1 _____   _____

    Hole #2 _____   _____

    Hole #3 _____   _____

    Hole #4 _____   _____

    (*see note under question 27 re calculator.) Is the part acceptable? _____.

29. What could be done to make the part good? _____

    _____

    _____

30. Referring to the following part and specified controls, fill in proper answers to each question.

   a. What is the datum hole position tolerance at MMC size? _____.

   b. What is the datum hole position tolerance at LMC size? _____.

   c. What is the datum hole perpendicularity tolerance at MMC size ? _____.

   d. What is the datum hole perpendicularity tolerance at LMC size? _____.

   e. What is the Ø.380 hole position tolerance at MMC size? _____.

   f. What is the virtual condition of the Ø.380 hole? _____.

   g. What is the virtual condition of the datum D hole? _____.

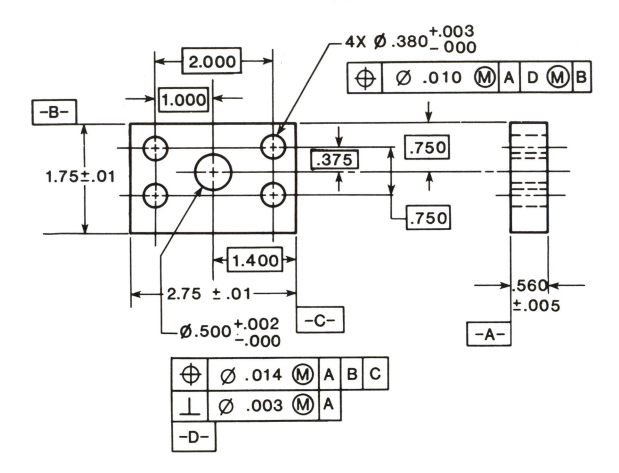

31. For what purpose do you believe the design required the perpendicularity control on the datum D hole on the preceding part? Select the answer below which most reflects your opinion.

   _____ Perfect form at MMC of related features is not controlled by size alone; therefore, perpendicularity was required to give such control.

   _____ The hole location could be more lenient but required a refinement of the perpendicularity to ensure part interface and proper assembly with a mating part.

32. If a functional gage were desired to verify the location of the datum D hole of figure shown under question 30 which datums would be picked up and in what precedence? _____, _____. Would MMC be applicable to the datums? _____. In the functional gage, what would the nominal size of the gage pin be? _____. What would the gage pin size be for the perpendicularity requirement? _____.

33. Functional gaging can also be used to evaluate the position of the Ø.380 holes and pattern under question 30. Add the nominal gage pin sizes to this illustration of a pin gage. (Disregard any consideration of gage tolerances).

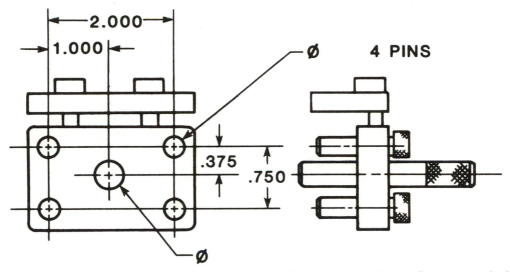

34. If a functional gage similar to that shown under question 33 were to be used on a part similar to that shown under question 30 - but where the datum is on an RFS basis and the features in control remain on an MMC basis - what difference would exist generally in the gage design and function? (Describe in words or by sketch below.) _____
_____.

35. Can position tolerancing be applied on an RFS basis? _____. Can an MMC position toleranced pattern of features be related to an RFS datum? _____. What must be added to the feature control symbol in the latter case? _____.

36. Complete the necessary details to ensure that the small part assembles to the larger one (via the tapped holes and clearance holes). The relationship of the hole pattern locations to the outside edges of each part is relatively unimportant (use .030); but in the pattern, the relationship and assigned tolerances must be calculated to ensure proper interface. Make the necessary determinations as to type of control required, make the calculations and complete the drawings below.

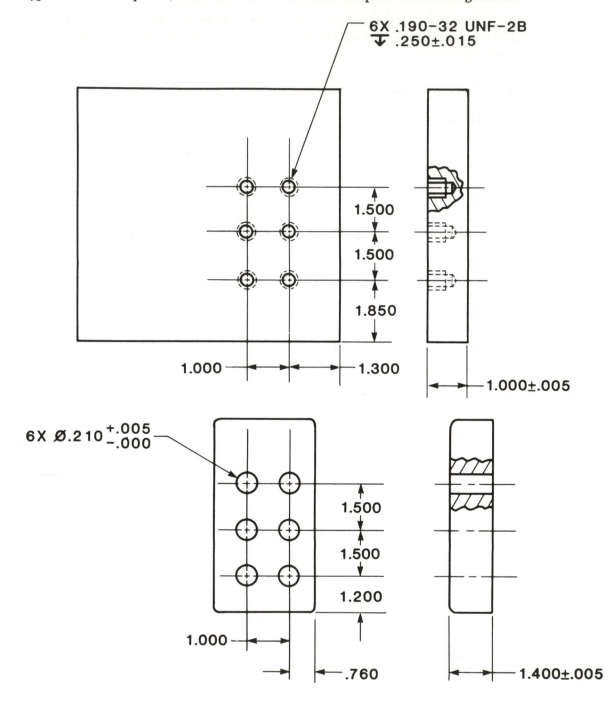

37. Design (sketch) a functional gage for the smaller part (one with clearance holes), to verify the in-the-pattern location controls only, as developed under question 36. You need only develop the nominal gage sizes, but show the gage construction as based upon your answer to question 36.

38. Datum targets have been partially shown on the part below. Noting the completed feature control frames and their specified datums, select the targets which appropriately construct these designated datum planes. Then, complete the datum target symbols and identify the targets according to your selection.

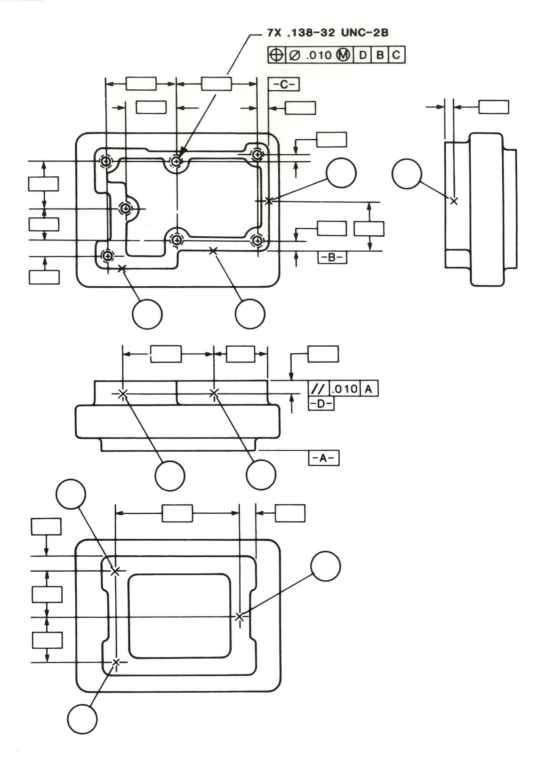

39. In the space below, design or sketch your own part using datum targets. Use a part similar to one from your own experience or establish an imaginary one. Establish the datum targets, datum planes, and show some feature relationships (e.g., hole pattern) with respect to the targets.

POSITION TOLERANCE ⊕—NON-CYLINDRICAL FEATURES—MATING PARTS

DESIGN CONSIDERATIONS

Position tolerancing is particularly practical and effective when controlling location of non-cylindrical mating part features on an MMC basis.

Where one part has slots and the mating part has external width features, the fixed fastener method of calculation can be used to determine the position tolerance on both part mating features. Symmetrical parts are typical of this application.

The sizes of the mating features are determined and specified as based upon the designer discretion and the design requirements.

The maximum material condition sizes of the mating features, i.e., the slot and related external width, are used to calculate the position tolerance for these features on both parts.

The results of the fixed fastener calculation establishes the positional tolerance for both parts. The formula (when S = Slot MMC and W = External MMC) is:

$$T = \frac{S - W}{2} \qquad .003 = \frac{.466 - .460}{2}$$

Where there is a relationship of only one feature to the datum feature on each part, an extension of the fixed fastener method may be used to directly derive maximum tolerance and yet assure function and assembly using the formula when D_1 = Datum Width Part #1 MMC and D_2 = Datum Slot Part #2 MMC.

$$T = \frac{(S - W) + (D_2 - D_1)}{2} \qquad .008 = \frac{(.466 - .460) + (.890 - .880)}{2}$$

Where desirable to select a more suitable distribution of tolerance between the mating part features, the calculated total tolerance may be divided between the parts (where .016 is the maximum total tolerance to be distributed, such combinations as .010 and .006, .011 and .005, etc.). This is done at the design stage before release to production.

Where maximum material condition is specified, the stated positional tolerances on each part are individually increased an amount equal to the departure from MMC size as the actual slots and widths are produced.

Corresponding features on both parts should be used as the datum references.

MEANING AND VERIFICATION PRINCIPLES

The clarity of the drawing requirements and the MMC principal are of advantage to production as they provide increased tolerance.

Functional gage principles may be utilized where the maximum material condition principles are specified. The gage member sizes are developed from each part to be gaged and their respective positional tolerances and slot and width sizes. Gage maker's tolerances must also be considered according to standard practices. It should be noted that the resulting virtual condition on each part determines the nominal gage member size.

POSITION TOLERANCE—NON-CYLINDRICAL FEATURES ⌖

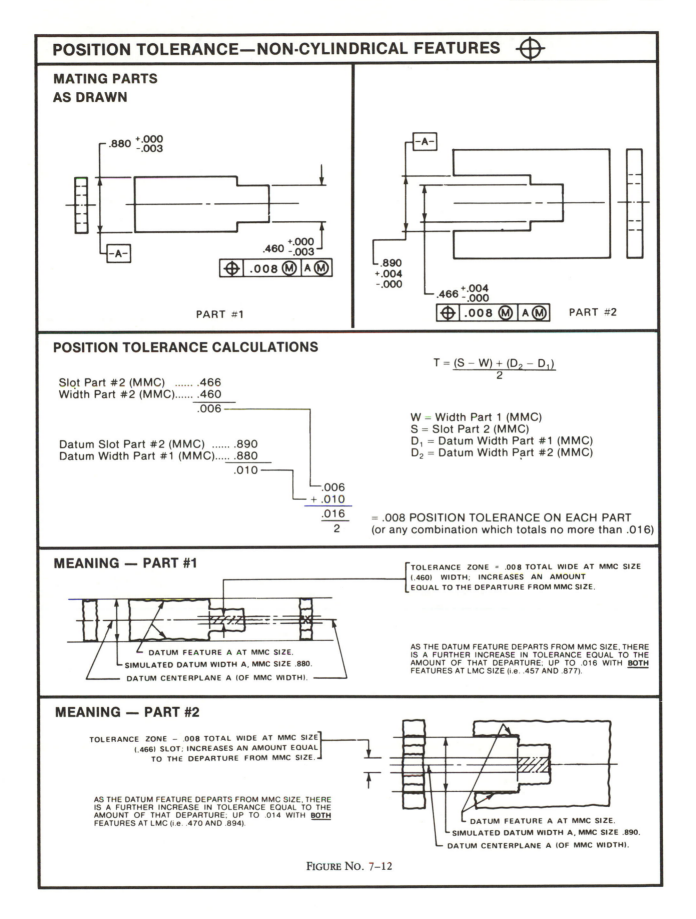

**MATING PARTS
AS DRAWN**

.880 +.000 / -.003

.460 +.000 / -.003

⌖ | .008 Ⓜ | A Ⓜ

PART #1

-A-

.890 +.004 / -.000

.466 +.004 / -.000

⌖ | .008 Ⓜ | A Ⓜ   PART #2

POSITION TOLERANCE CALCULATIONS

Slot Part #2 (MMC) .... .466
Width Part #2 (MMC)...... .460
          .006

Datum Slot Part #2 (MMC) ...... .890
Datum Width Part #1 (MMC)..... .880
          .010

          .006
        + .010
          .016
          ───
           2

$$T = \frac{(S - W) + (D_2 - D_1)}{2}$$

W = Width Part 1 (MMC)
S = Slot Part 2 (MMC)
D_1 = Datum Width Part #1 (MMC)
D_2 = Datum Width Part #2 (MMC)

= .008 POSITION TOLERANCE ON EACH PART
(or any combination which totals no more than .016)

MEANING — PART #1

TOLERANCE ZONE = .008 TOTAL WIDE AT MMC SIZE (.460) WIDTH; INCREASES AN AMOUNT EQUAL TO THE DEPARTURE FROM MMC SIZE.

DATUM FEATURE A AT MMC SIZE.
SIMULATED DATUM WIDTH A, MMC SIZE .880.
DATUM CENTERPLANE A (OF MMC WIDTH).

AS THE DATUM FEATURE DEPARTS FROM MMC SIZE, THERE IS A FURTHER INCREASE IN TOLERANCE EQUAL TO THE AMOUNT OF THAT DEPARTURE; UP TO .016 WITH **BOTH** FEATURES AT LMC SIZE (i.e. .457 AND .877).

MEANING — PART #2

TOLERANCE ZONE – .008 TOTAL WIDE AT MMC SIZE (.466) SLOT; INCREASES AN AMOUNT EQUAL TO THE DEPARTURE FROM MMC SIZE.

AS THE DATUM FEATURE DEPARTS FROM MMC SIZE, THERE IS A FURTHER INCREASE IN TOLERANCE EQUAL TO THE AMOUNT OF THAT DEPARTURE; UP TO .014 WITH **BOTH** FEATURES AT LMC (i.e. .470 AND .894).

DATUM FEATURE A AT MMC SIZE.
SIMULATED DATUM WIDTH A, MMC SIZE .890.
DATUM CENTERPLANE A (OF MMC WIDTH).

FIGURE NO. 7–12

POSTION TOLERANCE ⊕ —NON-CYLINDRICAL— GAGES

GAGE DESIGN CONSIDERATIONS

Position tolerances applied at MMC on non-cylindrical features permit the use of functional gages.

The use of functional gages simulate mating part interface and can expedite verification process.

The gage member sizes are developed from the MMC size and stated positional tolerance of the controlled feature. The virtual condition of the controlled feature is the nominal gage member size.

Gage maker's tolerances must also be applied according to standard practices.

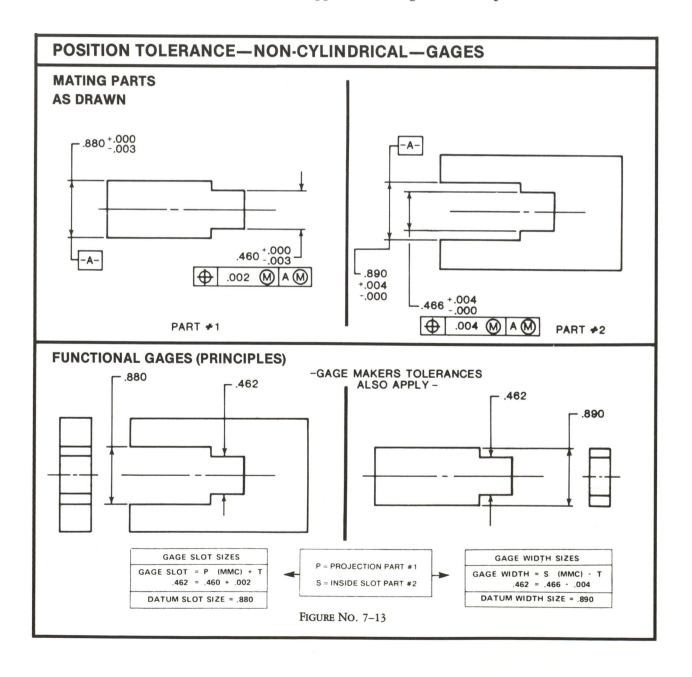

FIGURE NO. 7–13

RFS APPLICATION—DESIGN CONSIDERATIONS

Where higher precision parts of this variety (in this case, a symmetrical part) are required regardless of feature size (RFS), it is specified as shown on the part below. The RFS material condition symbol Ⓢ is placed in the feature control frame according to RULE #2—The Position Tolerance Rule. In this example, RFS is applied to both the feature controlled and the datum feature B. A stabilizing primary datum (A) is also used to establish the datum reference frame.

MEANING AND VERIFICATION PRINCIPLES

Note how the part is stabilized to the datums A and B. (The third plane longitudinal movement is irrelevant to the requirement). Also note the .010 wide position tolerance is maximum regardless of feature size.

Verification principles are shown. The derived center of the slot, when determined from the surfaces (the difference in the indicator readings), must be within .010. Other comparable methods are, of course, possible.

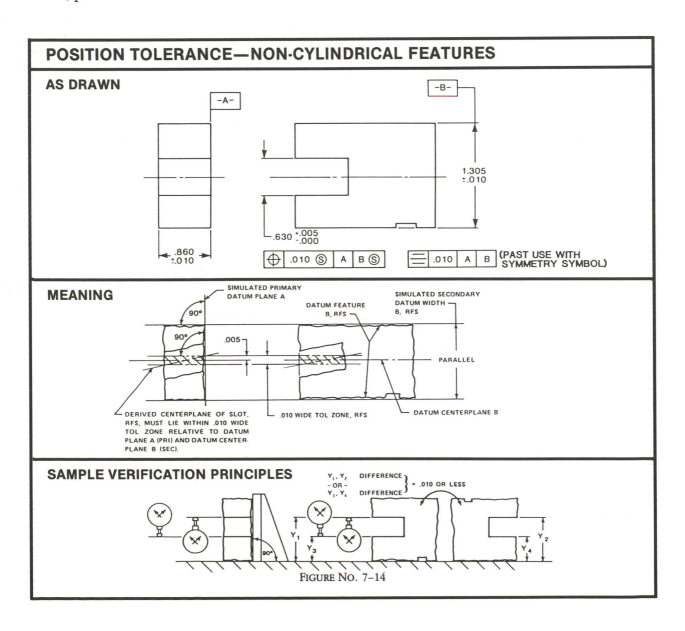

FIGURE NO. 7–14

POSITION TOLERANCE ⊕ —COAXIAL FEATURES— MATING PARTS

DESIGN CONSIDERATIONS

Position tolerancing is particularly practical and effective when controlling the location of coaxial mating part features on an MMC basis. See Fig. 7-15.

Where one part has clearance holes (bores, counterbores, etc.) and the mating part has corresponding features (pins, shafts, etc.), the fixed fastener method of calculation can be used to determine the position tolerances on both part mating features.

The sizes of the mating features are determined and specified as based upon the designer discretion and the design requirements.

The maximum material condition sizes of the mating features (the shaft and related) are used to calculate the position tolerance for these features on both parts.

The results of the fixed fastener calculation establishes the positional tolerance for both parts. The formula (when H = Hole MMC and S = Shaft MMC) is:

$$T = \frac{H - S}{2} \qquad .0025 = \frac{.711 - 706}{2}$$

Where there is a relationship of only one feature to the datum feature on each part, an extension of the fixed fastener method may be used to directly establish maximum tolerance and yet assure function and assembly. The formula (where D_1 = Datum Shaft MMC and D_2 = Datum Hole MMC) is:

$$T = \frac{(H - S) + (D_2 - D_1)}{2} \qquad .005 = \frac{(.711 - .706) + (.905 - .900)}{2}$$

Where desirable to select a more suitable distribution of tolerance between the mating part features, the calculated total tolerance may be divided between the parts (where .010 is the total tolerance to be distributed, such combinations as .006 and .004, .007 and .003, etc.). This is done at the design stage before release to production.

Where maximum material condition is specified, the stated positional tolerances on each part are individually increased an amount equal to the departure from MMC size as the actual holes and shafts are produced.

Corresponding features on both parts should be used as datum references.

MEANING AND VERIFICATION PRINCIPLES

The clarity of the drawing requirements and the advantages of the MMC principle with increased tolerance possibilities will assist production.

Functional gage principles may be utilized where the maximum material condition principles are specified. The gage member sizes are developed from each part to be gaged and their respective positional tolerances and hole and shaft sizes. Gage maker's tolerances must also be considered according to standard practices. It should be noted that the resulting virtual condition on each part determines the nominal gage member size.

POSITION TOLERANCE—COAXIAL FEATURES—MATING PARTS

AS DRAWN

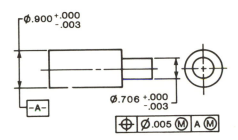

$\varnothing.900^{+.000}_{-.003}$

$\varnothing.706^{+.000}_{-.003}$

$\boxed{\oplus \ | \ \varnothing.005 \ Ⓜ \ | \ A \ Ⓜ}$

PART #1

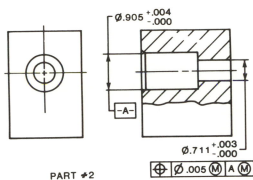

$\varnothing.905^{+.004}_{-.000}$

$\varnothing.711^{+.003}_{-.000}$

PART #2

$\boxed{\oplus \ | \ \varnothing.005 \ Ⓜ \ | \ A \ Ⓜ}$

POSITION TOLERANCE CALCULATIONS

```
HOLE PART #2, MMC . . . . . . . .  .711
SHAFT PART #1, MMC . . . . . . . (-).706
                                  ─────
                                   .005 ─┐
DATUM HOLE PART #2, MMC . . .     .905    │
DATUM SHAFT PART #1, MMC . . (-).900    ├─► .005
                                  ─────  │
                                   .005 ─┘ ─► (+) .005
                                             ─────────
```

$$T = \frac{(H - S) + (D_2 - D_1)}{2}$$

H = HOLE PART #2, MMC
S = SHAFT PART #1, MMC
D_1 = DATUM SHAFT PART #1, MMC
D_2 = DATUM HOLE PART #2, MMC

$\varnothing.005$ POSN TOL ON EACH PART
(OR ANY COMBINATION ON EACH PART WHICH TOTALS .010)

$$= \frac{.010}{2}$$

MEANING PART #1

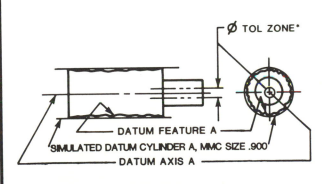

$\varnothing$ TOL ZONE*

DATUM FEATURE A

SIMULATED DATUM CYLINDER A, MMC SIZE .900

DATUM AXIS A

SHAFT LOCATION
THE AXIS OF THE SHAFT MUST LIE WITHIN TOL ZONE RELATIVE TO DATUM AXIS A.

*TOLERANCE ZONE = $\varnothing.005$ AT MMC SIZE (.706) OF SHAFT; INCREASES AN AMOUNT EQUAL TO THE DEPARTURE FROM MMC SIZE OF SHAFT; AND ADDITIONAL AMOUNT EQUAL TO THE DEPARTURE FROM MMC SIZE OF DATUM SHAFT UP TO $\varnothing.011$ MAX WITH BOTH FEATURES AT LMC SIZE (I.E., .703 & .897).

MEANING PART #2

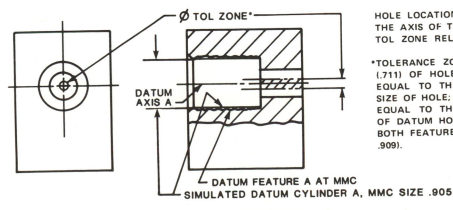

$\varnothing$ TOL ZONE*

DATUM AXIS A

DATUM FEATURE A AT MMC

SIMULATED DATUM CYLINDER A, MMC SIZE .905

HOLE LOCATION
THE AXIS OF THE HOLE MUST LIE WITHIN TOL ZONE RELATIVE TO DATUM AXIS A.

*TOLERANCE ZONE = $\varnothing.005$ AT MMC SIZE (.711) OF HOLE; INCREASES AN AMOUNT EQUAL TO THE DEPARTURE FROM MMC SIZE OF HOLE; AND ADDITIONAL AMOUNT EQUAL TO THE DEPARTURE FROM MMC OF DATUM HOLE UP TO $\varnothing.012$ MAX WITH BOTH FEATURES AT LMC SIZE (I.E., .714 & .909).

FIGURE NO. 7–15

POSITION TOLERANCE ⌖ —COAXIAL—GAGES

GAGE DESIGN CONSIDERATIONS

Positional tolerance applied to coaxial features on an MMC basis permits use of functional gages. Such gages simulate mating part relationships and facilitate verification techniques.

The gage member sizes are developed from the part features to be gaged, their MMC sizes, and the stated position tolerances (their virtual condition).

The formulas used are:

(GAGE HOLE = PART SHAFT MMC + POSITION TOLERANCE)

GH = S + T

(GAGE PIN (SHAFT) = PART HOLE MMC-POSITION TOLERANCE)

GP = H – T

Gage member sizes for the part datum features are established by their MMC sizes as stated on the drawing.

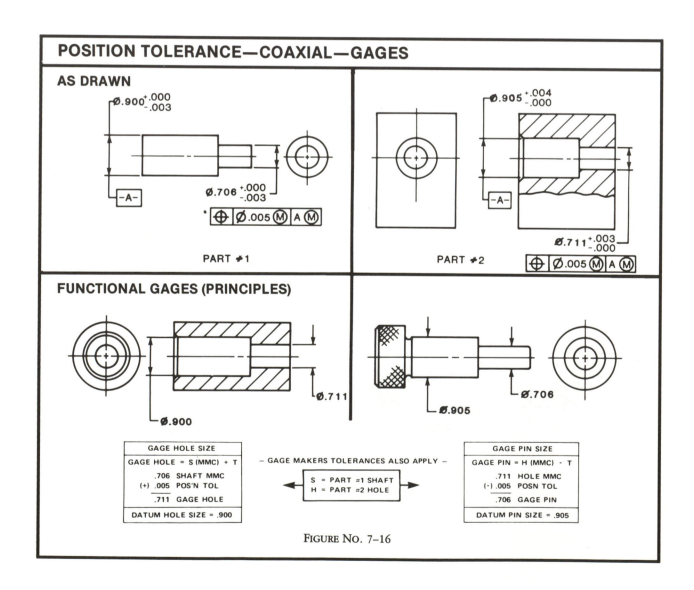

FIGURE NO. 7–16

POSITION TOLERANCE ⊕ — LEAST MATERIAL CONDITION Ⓛ

DEFINITION

Least material condition (LMC, Ⓛ) is the condition opposite to MMC, for example, the low limit of the stated size of a shaft or pin and the high limit of the stated hole size.

DESIGN CONSIDERATIONS

Least material condition provides a reverse method (from MMC) to apply position (and other) tolerances where exacting RFS principles are not warranted. It can provide an alternative method to controlling location (position) where the critical feature size limit is its least material condition at which size the stated tolerance applies. As the size of the controlled feature (a hole) departs from LMC (least material condition) the geometric tolerance increases.

Least material condition can be used to maintain critical center locations where MMC would be inappropriate, critical wall thickness relative to the controlled feature, compensating effect in precision location or alignment, etc.

MEANING

As seen in the drawing, the .005 position tolerance increases as the ∅.130 hole size as produced departs from LMC size toward MMC.

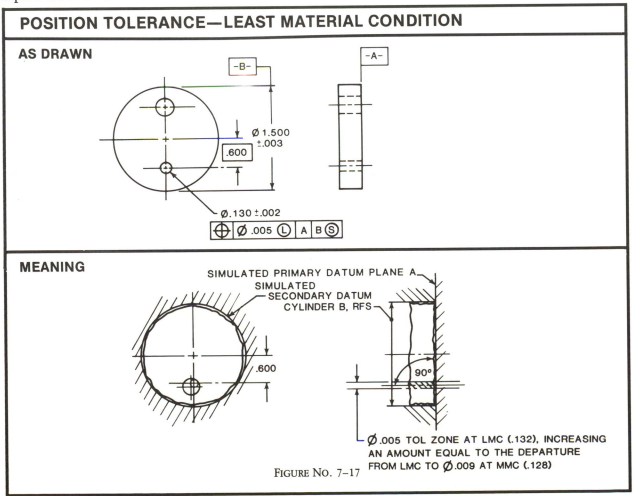

POSITION TOLERANCE—LEAST MATERIAL CONDITION

AS DRAWN

-B-

-A-

∅ 1.500 ±.003

.600

∅.130 ±.002

⊕ ∅ .005 Ⓛ A B Ⓢ

MEANING

SIMULATED PRIMARY DATUM PLANE A
SIMULATED SECONDARY DATUM CYLINDER B, RFS

.600

90°

∅ .005 TOL ZONE AT LMC (.132), INCREASING AN AMOUNT EQUAL TO THE DEPARTURE FROM LMC TO ∅ .009 AT MMC (.128)

FIGURE NO. 7–17

CONCENTRICITY ◎

DEFINITION

Concentricity is the condition where the axes of all cross-section elements of a feature's surface of revolution (cylinders, cones, spheres, hexagons, etc.) are common to the axis of a datum feature.

Concentricity tolerance is the diameter of the cylindrical tolerance zone within which the axis of the feature, or features, must lie. The axis of the tolerance zone must coincide with the axis of the datum feature or features.

DESIGN CONSIDERATIONS

Concentricity tolerance is an axis-to-axis type of control which can effectively relate coaxial features where part balance, uniform distribution of part feature mass in rotation, controlling the geometry of a non-rigid rotational part, etc. are required. See Fig. 7-18 for an example of application.

Large masses of material rotating at high speed about an axis, thin walled parts which distort under centrifugal forces, or precision axis-to-axis relationships where form is irrelevant to the functional axis, are examples of parts and features which might consider concentricity control necessary.

Concentricity tolerance is a more restrictive and potentially costly requirement due to the possible need for detailed analysis of the part in verification. Before concentricity tolerance is selected, the options of position tolerance at MMC or runout tolerance should be considered.

Concentricity tolerance considers in composite the effect of various surface error such as out-of-straightness, out-of-cylindricity, etc., as the resulting axis is determined.

Concentricity tolerance is always specified on an RFS basis.

MEANING AND VERIFICATION PRINCIPLES

Concentricity verification requires a form of differential measurements at opposed elements of the surface to determine the resulting feature axis from the detailed analysis. Where precision spindle machine methods are used, polargraph print-out and analysis with overlay gages will achieve the same results. Computerization analysis is also used where such capability is available.

All size tolerances must be met independent of the concentricity tolerance.

IF METRIC

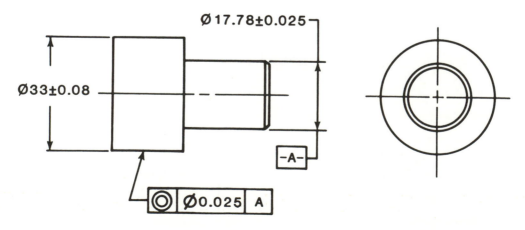

CONCENTRICITY ◎

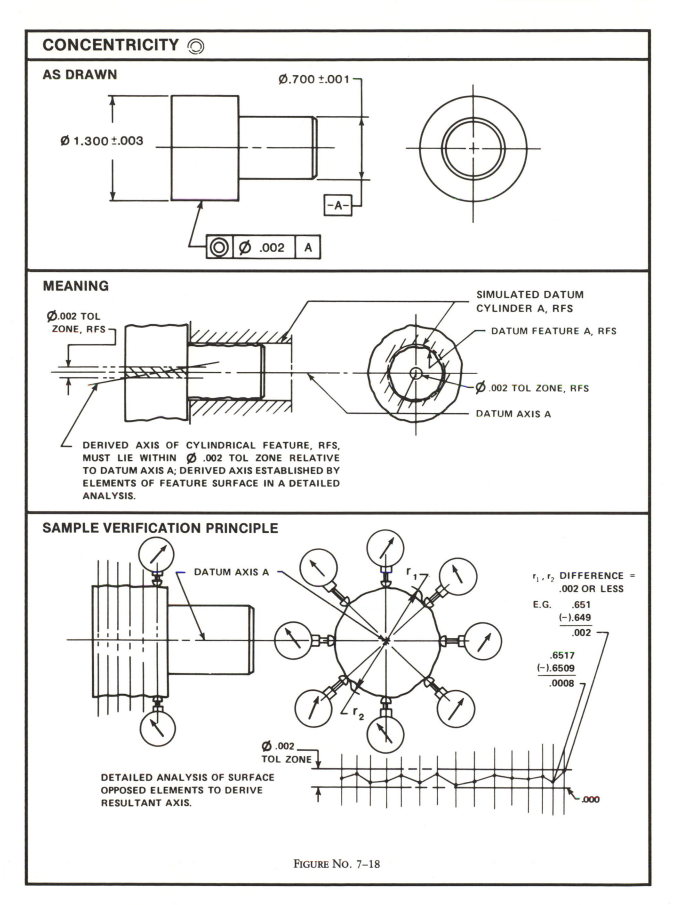

AS DRAWN

Ø.700 ±.001

Ø 1.300 ±.003

-A-

◎ | Ø | .002 | A

MEANING

Ø.002 TOL ZONE, RFS

SIMULATED DATUM CYLINDER A, RFS

DATUM FEATURE A, RFS

Ø .002 TOL ZONE, RFS

DATUM AXIS A

DERIVED AXIS OF CYLINDRICAL FEATURE, RFS, MUST LIE WITHIN Ø .002 TOL ZONE RELATIVE TO DATUM AXIS A; DERIVED AXIS ESTABLISHED BY ELEMENTS OF FEATURE SURFACE IN A DETAILED ANALYSIS.

SAMPLE VERIFICATION PRINCIPLE

DATUM AXIS A

r_1

r_2

r_1, r_2 DIFFERENCE = .002 OR LESS

E.G.   .651
    (−).649
        .002

    .6517
    (−).6509
        .0008

Ø .002 TOL ZONE

.000

DETAILED ANALYSIS OF SURFACE OPPOSED ELEMENTS TO DERIVE RESULTANT AXIS.

FIGURE NO. 7–18

7A
QUIZ-EXERCISES
POSITION TOLERANCE, NON-CYLINDRICAL FEATURES, COAXIAL FEATURES, LEAST MATERIAL CONDITION AND CONCENTRICITY

The following questions are relative to the material in this chapter. Read the question and answer to the best of your ability. The answers can be found in the companion manual *Answer Book and Instructor's Guide for Introduction to Geometric Dimensioning and Tolerancing.*

Consult your instructor if you have any questions.

1. Position tolerancing may be used on functional or assembly requirements of non-cylindrical features. On this part, specify that the .501 slot is to be located at true position (at MMC) with respect to the 1.120 width (at MMC) within .005 total tolerance.

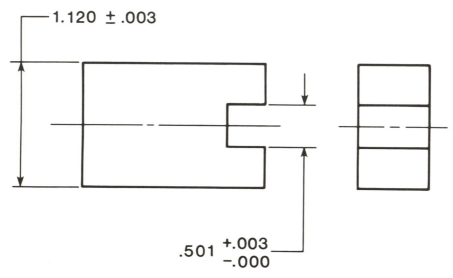

2. Sketch the part shown under question 1 and show the position tolerance zone.

3. Position tolerancing may be applied to relate non-cylindrical features of mating parts. Establish position tolerances on the mating parts shown below. Also calculate the maximum permissible production tolerance that could be permitted on each part if its feature and datum sizes were to depart from MMC size to least material size.

PART #1

PART #2

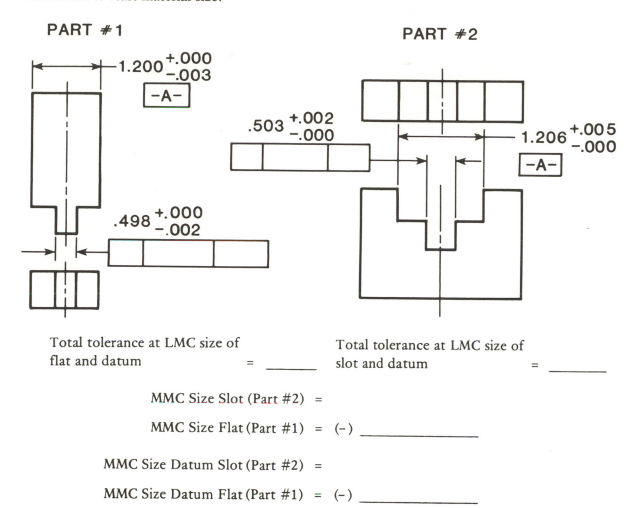

Total tolerance at LMC size of
flat and datum   = _____

Total tolerance at LMC size of
slot and datum   = _____

MMC Size Slot (Part #2) =

MMC Size Flat (Part #1) = (−) _____

MMC Size Datum Slot (Part #2) =

MMC Size Datum Flat (Part #1) = (−) _____

4. Position tolerancing may be used on functional or assembly requirements of coaxial features. On this part, specify that the ∅.305 diameter is to be located at true position (at MMC) with respect to the ∅.500 diameter (at MMC) within ∅.003 diameter tolerance zone.

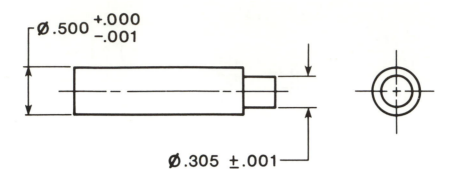

5. Sketch the part shown under question 4 and show the position tolerance zone.

6. Position tolerancing may be applied to relate coaxial features of mating parts. Establish position tolerances on the mating parts shown below. Also calculate the maximum permissible production tolerance that could be permitted on each part if its feature and datum sizes were to depart from MMC size to least material size.

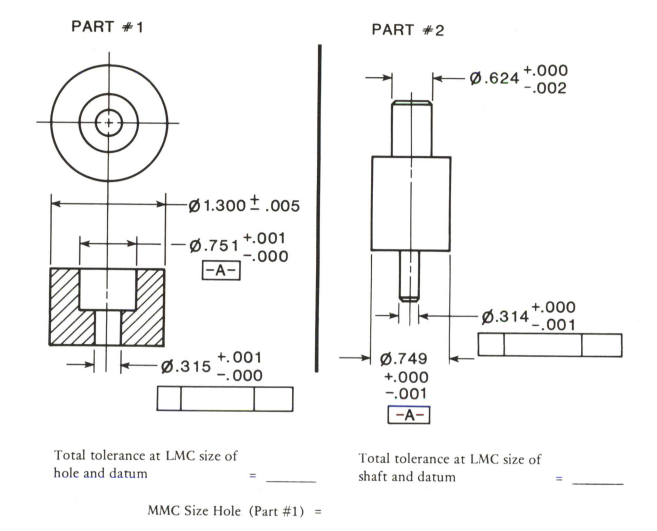

PART #1

Ø 1.300 ± .005

Ø .751 +.001 -.000

-A-

Ø .315 +.001 -.000

PART #2

Ø .624 +.000 -.002

Ø .314 +.000 -.001

Ø .749 +.000 -.001

-A-

Total tolerance at LMC size of
hole and datum   = _____

Total tolerance at LMC size of
shaft and datum   = _____

MMC Size Hole (Part #1) =

MMC Size Shaft (Part #2) = (−) _____

MMC Size Datum Hole (Part #1) =

MMC Size Datum Shaft (Part #2) = (−) _____

7. Where errors of form and position are considered on the basis of displacement of axis of two or more basically coaxial features, RFS, which control is used? _____. Establish necessary datums and complete the feature control frame for the part below. Assume that the two diameters ($\varnothing$.605 and $\varnothing$.500) are functional to the axis of rotation of the part with the 01.000 diameter relative to the resulting axis, within $\varnothing$.003 diameter tolerance, RFS.

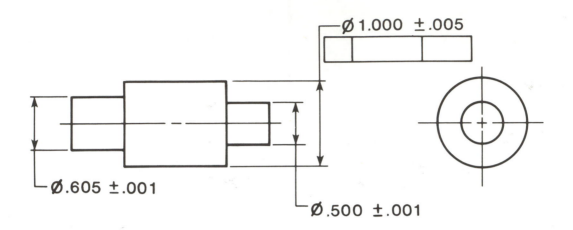

8. Using conventional FIM (FIR, TIR) methods of evaluation, if the part checked at .003 FIM, has it met the concentricity requirement? _____.

9. If the part error exceeds the stated tolerance when the FIM method is used, does this mean the part has not met the concentricity requirement? _____. Which statement supports your answer most appropriately?

_____ The surface may be out-of-round, etc., which will influence the reading but does not conclusively prove center (axis) displacement error.

_____ Concentricity is a variety of locational tolerance control, and the resulting error detected must be compared to the diameter tolerance zone.

10. Before a concentricity tolerance is specified, which other characteristics should be considered first if possible? _____ or _____. Which statement supports your answer most appropriately?

_____ Concentricity requirements are encountered less frequently.

_____ MMC methods or conventional surface criteria controls are more readily producible and economical.

11. Sketch the part shown under question 7 and show the concentricity tolerance zone.

12. This part has symmetrical features. Specify that the .501 width of this part is to be related to the 1.120 width within .005 total tolerance, RFS.

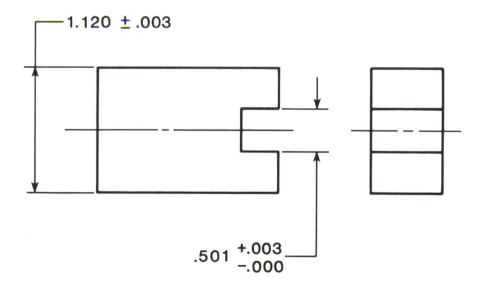

1.120 ± .003

.501 +.003
     −.000

13. Sketch the part shown under question 12 and show the tolerance zone.

14. Indicate on the below part that the 4 holes are located at true position within .010 diameter at LMC with respect to datum A (the bottom face of the part). Disregard the pattern location from the outside edges for purposes of this example.

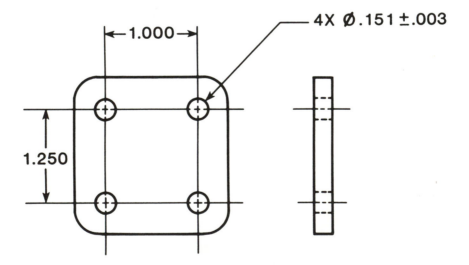

8
SUPPLEMENTAL INFORMATION

THE FEATURE CONTROL FRAME

The Feature Control Frame is:

What the Feature Control Frame Contains

The feature control frame comprises the pictorial "note" which includes:

        a. Kind of control (geometric characteristic),

        b. The geometric tolerance,

        c. Any modifiers (i.e., Ⓜ, Ⓢ, Ⓛ),

        d. Datum references, and any datum reference modifiers (i.e., Ⓜ, Ⓢ, Ⓛ).

Arrangement of the Datum References in the Feature Control Frame

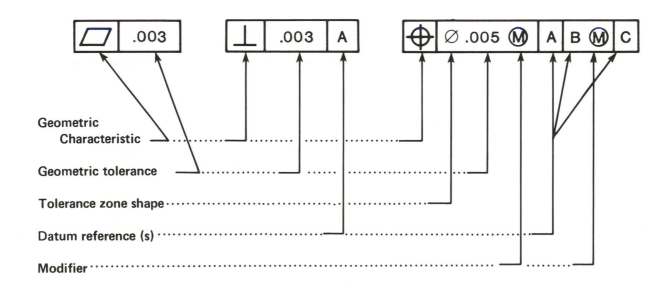

Datum Precedence Signified by Placement of Datum Reference Letters

Reading left to right, the datum reference letters indicate an order of precedence of the datum features so identified. There is no significance of the alphabetic sequence.

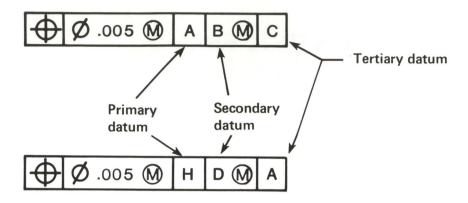

Multiple Datum Reference Letters to Indicate a Common Datum

The use of two datum letters separated by a dash indicates a common datum (i.e., axis or center-plane) is established by two datum features; therefore, there is no precedence between the two features so identified; together they create a common datum.

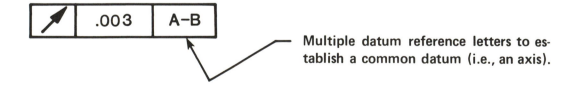

Multiple datum reference letters to establish a common datum (i.e., an axis).

Placement of the Feature Control Frame on the Drawing

The feature control frame is placed on the drawing according to standard drawing conventions and principles. Some typical applications are:

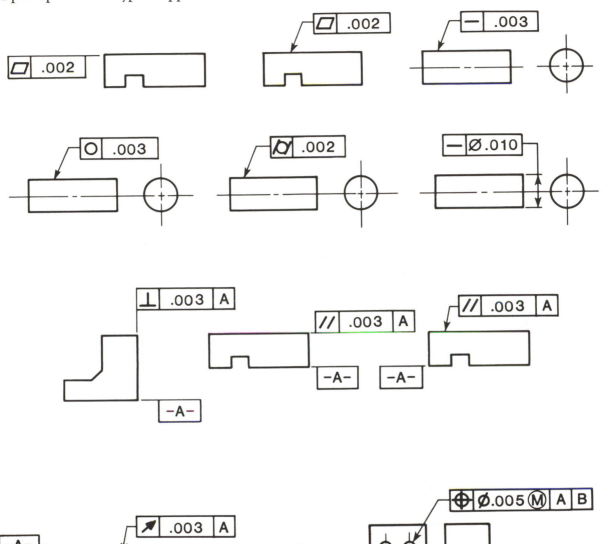

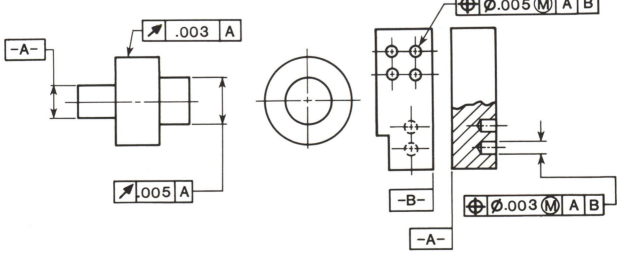

THE DATUM FEATURE SYMBOL

The datum feature symbol specifies the feature(s) of a part from which functional relationships are established. A geometric tolerance is then indicated with respect to that datum feature.

The Datum Feature Symbol is:

Datum Reference Letter

Each feature requiring identification as a datum on a drawing uses a difference reference letter.

Any letters except I, O, or Q may be used (these letters would cause some confusion as they resemble numbers or other symbols). It is common to use the first three letters A, B, C of the alphabet first on a drawing, although it is not required, nor does the choice of letters have any significance. Where the alphabet is exhausted, double letters such as AA, AB, etc. may be used.

The dash lines are to make the letter stand out in the symbol, and make the symbol distinctive from other drawing conventions. The lines have no other significance.

Placement of the Datum Feature Symbol on the Drawing

The datum identification symbol is placed on the drawing according to standard drawing conventions. Specific significance is indicated, however, by certain placement. Some typical applications are:

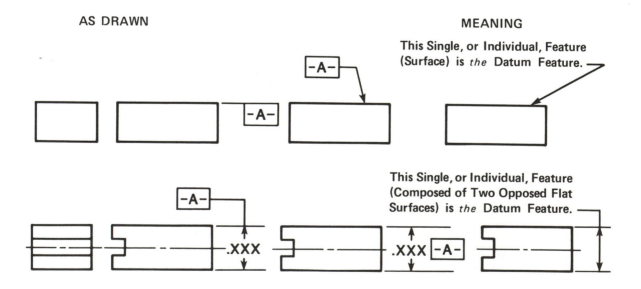

AS DRAWN

MEANING

This Single, or Individual, Feature (the Cylinder) is *the* Datum Feature.

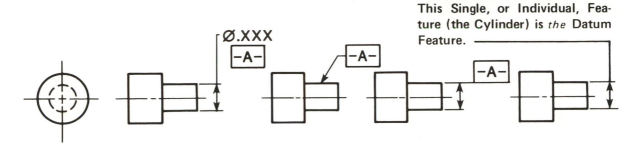

This Single, or Individual, Feature (the Cylinder) is *the* Datum Feature.

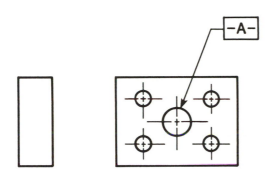

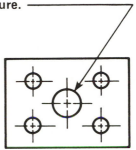

COMBINATION FEATURE CONTROL FRAME AND DATUM FEATURE SYMBOL

For readability of the drawing, consolidating information, conservation of drawing space, and convenience, the Feature Control Frame and Datum Feature Symbol may be combined on a drawing.

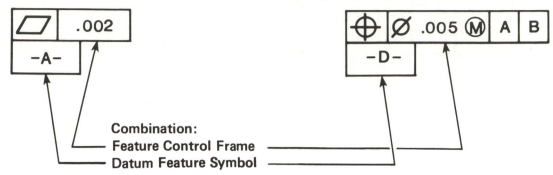

Combination:
Feature Control Frame
Datum Feature Symbol

Combining, or attaching, one symbol to the other has no technical significance. The feature control frame gives the indicated feature its control; the datum feature symbol its identity (name).

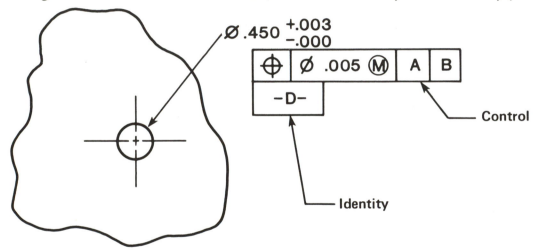

Combined or separately stated, the symbols have the same meaning.

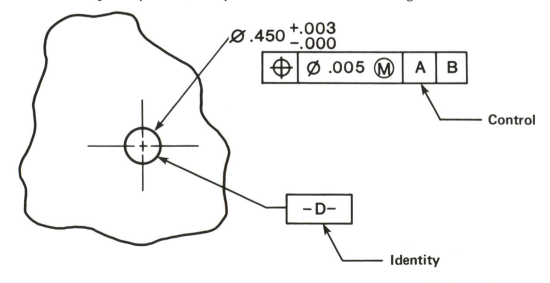

For further convenience, although not encouraged, the datum identification symbol length may be extended to that of the feature control symbol.

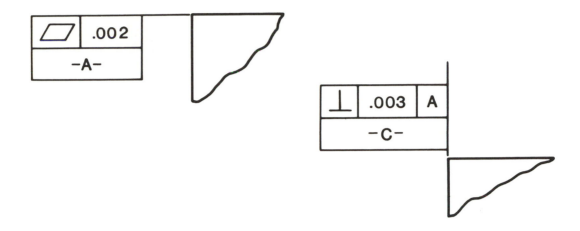

There is no significance to the placement of the Feature Control Frame and the Datum Feature Symbol with respect to one another (above or below, etc.).

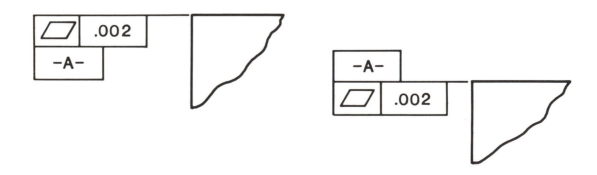

THE STANDARD RULES

The rules of geometrics provide all users of the system a common bridge of understanding on certain key fundamentals; they provide an authoritative agreement of principle as based upon the established standards authority (ANSI Y14.5M-1982).

The rules provide ''handles'' for the users to better understand the system and also as a foundation for detailed application.

The rules emphasize and utilize the important principle of standards and standardization without which there could be no system. Through the voluntary consensus process of developing standards, such things as experience, history, technical improvements, needs (present and future), etc. are considered and debated to develop common ground agreements (standards). The rules emphasize and fix, for all users, specific ground rules on these key fundamentals.

Learning these fundamentals and the purposes of the rules will greatly assist in applying the system correctly. More confidence and effective application will result.

Relationships of "Size" to "Geometric Tolerancing"

Early in this text, the emphasis was placed on the importance of size controls as one essential in specifying and achieving design requirements. However, the limitations of size control was also emphasized; i.e., size alone does not control relationships of orientation, runout, or location. Where the standard is invoked as the basis for the drawing authority, this point is made clear. However, this same authority also assures that size does control form of individual features. These points are extremely important to universal understanding. The rules single-out and emphasize these key fundamentals and their meaning.

RULE #1 · LIMITS OF SIZE RULE

Individual Features of Size

Where only a tolerance of size is specified, the limits of size of the individual feature prescribe the extent to which variations in its geometric form as well as size are allowed.

Variations of Form

The form of an individual feature is controlled by its limits of size to the extent prescribed in the following paragraph and illustration:

a. The surface, or surfaces, of a feature shall not extend beyond a boundary (envelope) of perfect form at MMC. This boundary is the true geometric form represented by the drawing. No variation is permitted if the feature is produced at its MMC limit of size.

INDIVIDUAL SIZE FEATURE

Rule #1, continued

    b. Where the actual size of a feature has departed from MMC toward LMC, a variation in form is allowed equal to the amount of such departure.

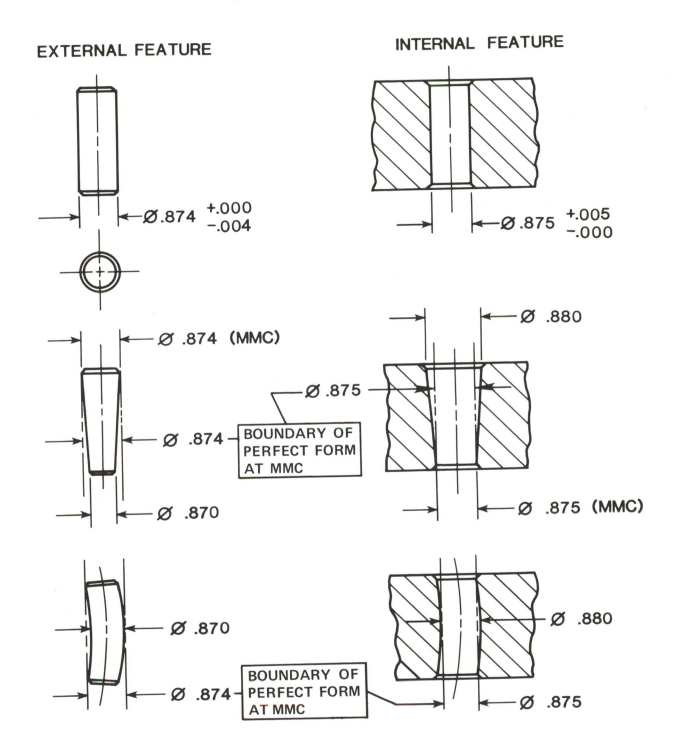

Rule #1, continued

c. The actual size of an individual feature at any cross section shall be within the specified tolerance of size.

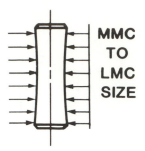

Relationship Between Individual Features

The limits of size do not control the orientation, runout or location relationship between individual features. Such features, if shown perpendicular, coaxial, or otherwise geometrically related to each other, must be controlled to avoid incomplete drawing requirements. The use of orientation, runout or location tolerances are then necessary.

To Establish Boundary of Perfect Form at MMC Between Features

If it is desired to establish a boundary of perfect form at MMC to control the relationship between features, the following methods may be used:

a. Specify a zero tolerance of orientation or positional tolerance at MMC, including necessary datum references.

b. Indicate this control for the features involved by a note such as ''PERFECT (ORIENTATION, COAXIALITY, OR SYMMETRY) AT MMC REQUIRED FOR RELATED FEATURES.''

Perfect Form at MMC Not Required

Where it is desired to permit a surface, or surfaces, of a feature to exceed the boundary of perfect form at MMC, a note such as PERFECT AT MMC NOT REQ'D is specified exempting the pertinent size dimension from the provision of Rule #1.

The control of geometric form prescribed by limits of size does not apply to the following:

a. Stock such as bars, sheets, tubing, structural shapes, and other items produced to establish industry or government standards which prescribe limits for straightness, flatness, and other geometric characteristics. Unless geometric tolerances are specified on the drawing of a part made from these items, standards for the items govern the surfaces that remain in the ''as furnished'' condition on the finished part.

b. Parts subject to free state variation in the unrestrained condition.

RULE #2 · POSITION TOLERANCE RULE

For a tolerance of position, Ⓜ, Ⓢ, or Ⓛ must be specified on the drawing with respect to the individual tolerance, datum reference, or both, as applicable.

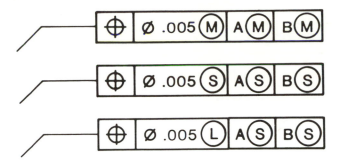

Former practice (Rule #2)*

For a tolerance of position, MMC applies with respect to an individual tolerance, datum reference, or both, where no condition is specified. RFS must be specified where it is required.

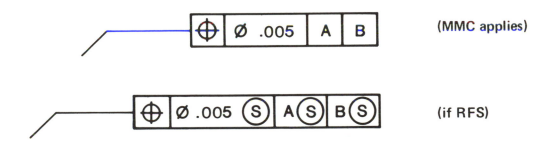

Note: Former practice (Rule #2) not recommended for new application.*

*Retained for information only to interpret drawings prepared under earlier standards (i.e., Y14.5-1973).

RULE #3 - OTHER THAN POSITION TOLERANCE RULE

For all applicable geometric tolerances, other than position tolerance, RFS applies with respect to the individual tolerance, datum reference, or both, where no modifying symbol is specified. Ⓜ must be specified on the drawing where it is required.

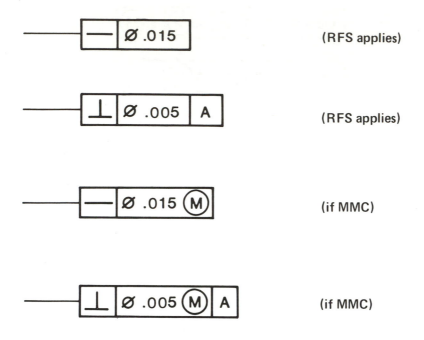

⎯ | Ø .015	(RFS applies)
⊥ | Ø .005 | A	(RFS applies)
⎯ | Ø .015 Ⓜ	(if MMC)
⊥ | Ø .005 Ⓜ | A	(if MMC)

Characteristics and controls which can be applicable to size features and, thus, to which RFS applies under Rule #3 unless modified to MMC, are:

Characteristics and controls which are always applicable at RFS under Rule #3 and, due to the nature of the requirement, cannot be applied at MMC, are:

*Use POSITION instead per ANSI Y14.5M-1982.

PITCH DIAMETER RULE*

Each tolerance or orientation or position and datum reference specified for a screw thread applies to the axis of the thread derived from the pitch cylinder.

(Means: Applies to the Pitch Diameter)

Where an exception to this practice is necessary, the specific feature of the screw thread (such as MAJOR Ø or MINOR Ø) shall be stated beneath the feature control frame or beneath the datum feature symbol, as applicable.

Each tolerance of orientation or location and datum reference specified for gears, splines, etc. must designate the specific feature of the gear, spline, etc. to which it applies (such as PITCH Ø, PD, MAJOR Ø, or MINOR Ø). This information is stated beneath the feature control frame or beneath the datum feature symbol.

*Formerly called Rule #4 of ANSI Y14.5-1973).

DATUM/VIRTUAL CONDITION RULE*

Depending on whether it is used as a primary, secondary, or tertiary datum, a virtual condition exists for a datum feature of size where its axis or centerplane is controlled by a geometric tolerance. In such a case, the datum feature applies as its virtual condition even though it is referenced in a feature control at MMC.

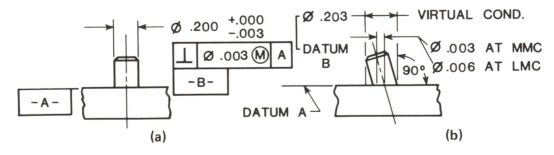

If the Ø.200 pin is established as datum feature B (Fig. (a) above), and is subsequently used in a datum reference frame along with datum feature A (the surface) to establish further feature relationships from both A and B (i.e., ⊕ | Ø.005 Ⓜ | A | B Ⓜ), the virtual condition of datum B is the basis for the relationship not MMC condition (Fig. (b) above).

*Formerly called Rule #5 of ANSI Y14.5-1973).

MATERIAL CONDITIONS (MODIFIERS)

MAXIMUM MATERIAL CONDITION

Symbol: Ⓜ

Abbreviation: MMC

Definition: The condition where a feature of size contains the maximum amount of material within the stated limits of size; for example, minimum hole diameter and maximum shaft diameter.

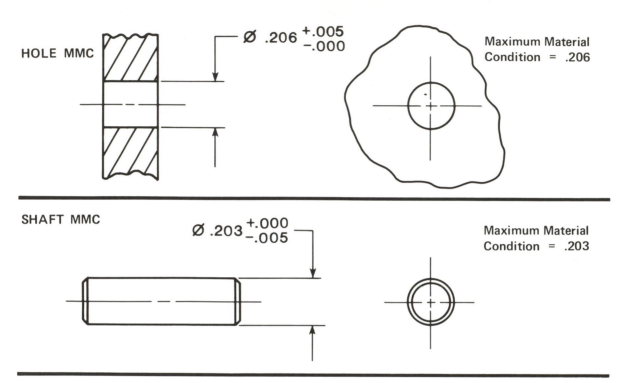

Maximum Material Condition can be applicable only to size features, such as holes, shafts, pins, slots, etc.; features which have an axis, centerline, or centerplane. MMC cannot be applied, for example, to a surface (plane) feature; a surface has no size as it is recognized in the geometrics system; it has area, but no third dimension nor axis, centerline, or centerplane.

Maximum Material Condition as a Concept

Maximum Material Condition is defined in the above text and illustrations insofar as it relates to size. However, MMC's real reason for existence is as a concept for design and manufacture. It captures the functional aspects of part features and relationships of mating features such as on mating parts and where interchangeability is required; it interrelates size with form, orientation, or position tolerances.

Therefore, there is always some type of relationship of size features to make MMC application valid. However, it is also in this prerequisite that the almost unlimited potential of the MMC concept or principle can be seen.

Please see applications in preceeding text for application of the MMC concept.

REGARDLESS OF FEATURE SIZE

Symbol: Ⓢ

Abbreviation: RFS

Definition: The geometrical tolerance applies at any increment of size of the feature within its size tolerance.

Regardless of Feature Size can be applied only to size features, such as holes, shafts, pins, slots, etc.,—features which have an axis, centerline, or centerplane. RFS is not applicable to a surface (plane) feature, for example.

The Regardless of Feature Size principle is invoked automatically by Rule #3 when a geometric tolerance (other than Position) is applied.

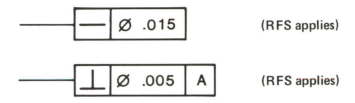

(RFS applies)

(RFS applies)

When position tolerance is applied and the RFS principle is desired, it must be stated by the RFS modifier (according to Rule #2).

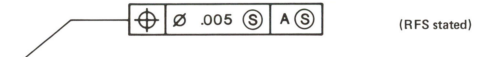

(RFS stated)

LEAST MATERIAL CONDITION

Symbol: Ⓛ

Abbreviation: LMC

Definition: The condition where a feature of size contains the least amount of material within the stated limits of size; for example, maximum hole diameter, minimum shaft diameter.

Least Material Condition can be applied only to size features, such as holes, shafts, pins, slots, etc.; features which have an axis, centerline, or centerplane.

When Position Tolerance is applied, the LMC principle is desired, it must be stated by the LMC modifier (according to Rule #2).

(LMC stated)

TOLERANCE ZONES

The Tolerance Zone describes numerically, as well as pictorially represents, the extent of the permissible deviation from the desired form, orientation, profile, runout, or location of the controlled feature. The tolerance zone may apply to a surface, axis, or centerplane.

In applying geometrics, all tolerances (i.e., the Tolerance Zones) which are shown in the feature control frame are total values. They describe the size and shape of the Tolerance Zone as based upon the kind of control specified.

Shape of Tolerance Zone

The tolerance zone either takes the shape relative to the kind of control (characteristic) used (i.e., flatness relates to a plane) automatically, is in the direction as described by the dimensional construction (a total wide zone), or by use of the diameter symbol (Ø) (a cylindrical zone).

Where the specified tolerance value is to indicate the diameter of a cylindrical zone, the diameter symbol (Ø) is placed in the feature control frame as shown below.

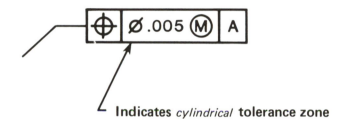

Indicates *cylindrical* tolerance zone

In all other applications, the tolerance zone applies and represents the distance between two parallel lines, planes, or geometric boundaries in the direction indicated or as understood by the geometrical control, such as below:

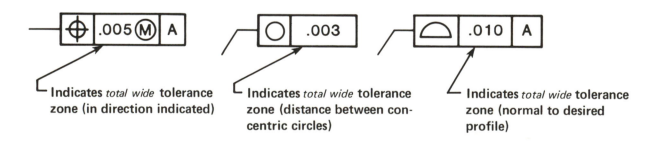

Indicates *total wide* tolerance zone (in direction indicated)

Indicates *total wide* tolerance zone (distance between concentric circles)

Indicates *total wide* tolerance zone (normal to desired profile)

Please see appropriate sections in the preceding text for more detail relative to each geometric control.

CONVERSION OF POSITION TOLERANCE ZONE ⊕ TO/FROM COORDINATE TOLERANCE ZONE

Conversion from the stated positional tolerance to the equivalent coordinate (±) tolerance or from the coordinate (±) to positional tolerance can be quickly derived from the chart.

This method does not provide assured precision; its use is, thus, to be handled with discretion.

The chart is entered with either the position or ± tolerance to derive the opposite value. The sample application explains the methods used.

The chart method may also be used to convert differentials in coordinate (X and Y) measurement of positionally toleranced features. That is, for example, if the measured coordinate results were:

If Hole Actual Location: (Horizontal X)   Basic (on Drawing) − Actual = X
.900 − .896 = .004 (In X Direction)

(Vertical Y)   Basic (on Drawing) − Actual = Y
.800 − .798 = .002 (In Y Direction)

Then Enter Chart on ± Scale in X and Y Directions.   .004 in X (Enter at ±.004 Line)
.002 in Y (Enter at ±.002 Line)   = .009 ⊕ Tol. as converted.

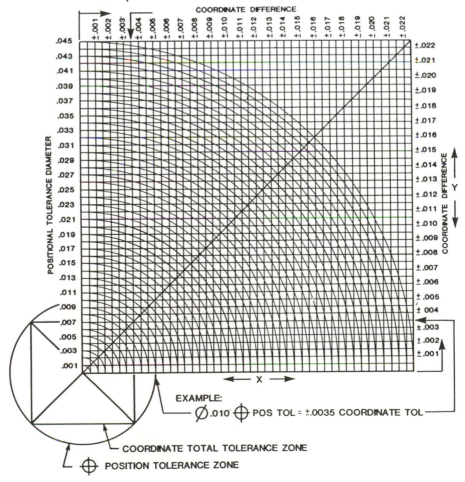

EXAMPLE:
Ø.010 ⊕ POS TOL = ±.0035 COORDINATE TOL

CONVERSION OF POSITION TOLERANCE ZONE ⊕ TO/FROM COORDINATE TOLERANCE ZONE— RULE OF THUMB

Conversion from the stated positional tolerance on the drawing to equivalent ± tolerances may be necessary for tool building, prototype parts manufacture, inspection, etc.

Tool designers, tool makers, machinists, model makers, inspectors, etc. can convert positional tolerances to equivalent ± tolerances by use of the "Rule of Thumb."

Conversion from the stated coordinate (±) tolerance to the equivalent positional tolerance can be useful to production engineers, inspectors, etc. who may wish to isolate possible problem areas; such as, where parts may assemble but have been previously rejected on the basis of the permissible coordinate tolerance on the drawing. This method may help "trouble-shoot" problems in general.

This is not the method used to determine positional tolerances in design. Proper methods of calculation are found elsewhere in this text.

Note that the ± tolerance to position tolerance conversion should be based upon the coordinate (±) tolerance allotted to the one hole (or other feature) under consideration. This usually requires basing the calculation upon one-half the stated coordinate tolerance between the holes (or other features).

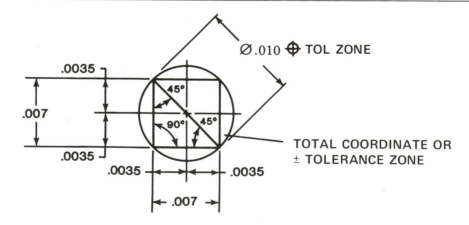

⊕ TO ±

⊕ TOL ZONE x .70711 = TOTAL ± TOL ZONE (FOR FEATURE)

EXAMPLE: ∅.010 ⊕ TOL x .70711 = .0071 → .007 TOTAL COORDINATE TOL OR ± .0035

<u>RULE OF THUMB:</u>
USE .7 (OR 70%) OF TOTAL ⊕ TOL TO CONVERT IN NON-CRITICAL APPLICATIONS, E.G., .7 x .010 = .007 (± .0035)

∅.010 ⊕ TOL ZONE

.0035
.007
.0035
45°
90°
45°
.0035   .0035
.007

TOTAL COORDINATE OR ± TOLERANCE ZONE

± TO ⊕

TOTAL ± ZONE x 1.4142 = ⊕ TOL ZONE

EXAMPLE: .007 TOTAL COORDINATE TOL
OR
± .0035
} x 1.4142 = ∅ .0099 OR ⊕ ∅ .010 TOL

<u>RULE OF THUMB:</u>
USE 1.4 TIMES TOTAL ± TOL TO CONVERT IN NON-CRITICAL APPLICATIONS, E.G., 1.4 x .007 = ∅ .010

CONVERSION OF COORDINATE MEASUREMENTS ±
TO POSITION LOCATION ⊕

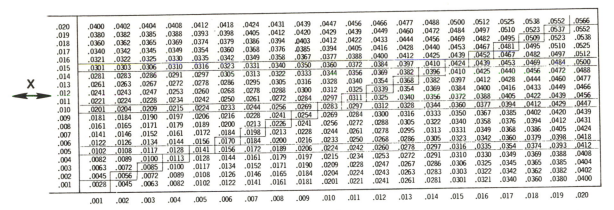

X \ Y	.001	.002	.003	.004	.005	.006	.007	.008	.009	.010	.011	.012	.013	.014	.015	.016	.017	.018	.019	.020
.020	.0400	.0402	.0404	.0408	.0412	.0418	.0424	.0431	.0439	.0447	.0456	.0466	.0477	.0488	.0500	.0512	.0525	.0538	.0552	.0566
.019	.0380	.0382	.0385	.0388	.0393	.0398	.0405	.0412	.0420	.0429	.0439	.0449	.0460	.0472	.0484	.0497	.0510	.0523	.0537	.0552
.018	.0360	.0362	.0365	.0369	.0374	.0379	.0386	.0394	.0403	.0412	.0422	.0433	.0444	.0456	.0469	.0482	.0495	.0509	.0523	.0538
.017	.0340	.0342	.0345	.0349	.0354	.0360	.0368	.0376	.0385	.0394	.0405	.0416	.0428	.0440	.0453	.0467	.0481	.0495	.0510	.0525
.016	.0321	.0322	.0325	.0330	.0335	.0342	.0349	.0358	.0367	.0377	.0388	.0400	.0412	.0425	.0439	.0452	.0467	.0482	.0497	.0512
.015	.0301	.0303	.0306	.0310	.0316	.0323	.0331	.0340	.0350	.0360	.0372	.0384	.0397	.0410	.0424	.0439	.0453	.0469	.0484	.0500
.014	.0281	.0283	.0286	.0291	.0297	.0305	.0313	.0322	.0333	.0344	.0356	.0369	.0382	.0396	.0410	.0425	.0440	.0456	.0472	.0488
.013	.0261	.0263	.0267	.0272	.0278	.0286	.0295	.0305	.0316	.0328	.0340	.0354	.0368	.0382	.0397	.0412	.0428	.0444	.0460	.0477
.012	.0241	.0243	.0247	.0253	.0260	.0268	.0278	.0288	.0300	.0312	.0325	.0339	.0354	.0369	.0384	.0400	.0416	.0433	.0449	.0466
.011	.0221	.0224	.0228	.0234	.0242	.0250	.0261	.0272	.0284	.0297	.0311	.0325	.0340	.0356	.0372	.0388	.0405	.0422	.0439	.0456
.010	.0201	.0204	.0209	.0215	.0224	.0233	.0244	.0256	.0269	.0283	.0297	.0312	.0328	.0344	.0360	.0377	.0394	.0412	.0429	.0447
.009	.0181	.0184	.0190	.0197	.0206	.0216	.0228	.0241	.0254	.0269	.0284	.0300	.0316	.0333	.0350	.0367	.0385	.0402	.0420	.0439
.008	.0161	.0165	.0171	.0179	.0189	.0200	.0213	.0226	.0241	.0256	.0272	.0288	.0305	.0322	.0340	.0358	.0376	.0394	.0412	.0431
.007	.0141	.0146	.0152	.0161	.0172	.0184	.0198	.0213	.0228	.0244	.0261	.0278	.0295	.0313	.0331	.0349	.0368	.0386	.0405	.0424
.006	.0122	.0126	.0134	.0144	.0156	.0170	.0184	.0200	.0216	.0233	.0250	.0268	.0286	.0305	.0323	.0342	.0360	.0379	.0398	.0418
.005	.0102	.0108	.0117	.0128	.0141	.0156	.0172	.0189	.0206	.0224	.0242	.0260	.0278	.0297	.0316	.0335	.0354	.0374	.0393	.0412
.004	.0082	.0089	.0100	.0113	.0128	.0144	.0161	.0179	.0197	.0215	.0234	.0253	.0272	.0291	.0310	.0330	.0349	.0369	.0388	.0408
.003	.0063	.0072	.0085	.0100	.0117	.0134	.0152	.0171	.0190	.0209	.0228	.0247	.0267	.0286	.0306	.0325	.0345	.0365	.0385	.0404
.002	.0045	.0056	.0072	.0089	.0108	.0126	.0146	.0165	.0184	.0204	.0224	.0243	.0263	.0283	.0303	.0322	.0342	.0362	.0382	.0402
.001	.0028	.0045	.0063	.0082	.0102	.0122	.0141	.0161	.0181	.0201	.0221	.0241	.0261	.0281	.0301	.0321	.0340	.0360	.0380	.0400

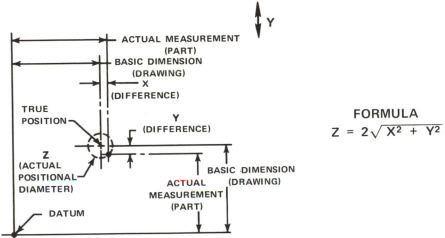

FORMULA

$$Z = 2\sqrt{X^2 + Y^2}$$

Part is set up on datum features (using datum precedence) to establish X and Y measuring planes.

Holes are measures in X and Y, reading and recording the results.

Holes are measured using RFS invoking MMC principles when necessary.

Measurement differentials (from basic to actual) are determined.

Chart (or other comparable methods) converts the differentials to equivalent positional values which is compared to the stated permissible positional tolerance.

Where necessary, the hole actual size is determined to invoke the MMC principle and derive permissible tolerance for that hole.

Coordinate measuring machines with computer capability, special computer software or programmable calculators can be utilized to perform such calculations.

COORDINATE MEASUREMENTS TO POSITION ⊕ LOCATION

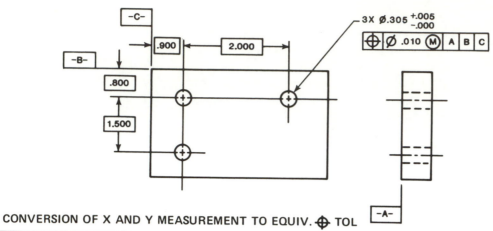

CONVERSION OF X AND Y MEASUREMENT TO EQUIV. ⊕ TOL

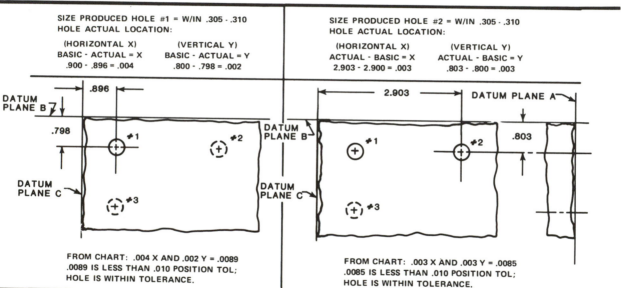

SIZE PRODUCED HOLE #1 = W/IN .305 - .310
HOLE ACTUAL LOCATION:

(HORIZONTAL X)	(VERTICAL Y)
BASIC - ACTUAL = X	BASIC - ACTUAL = Y
.900 - .896 = .004	.800 - .798 = .002

FROM CHART: .004 X AND .002 Y = .0089
.0089 IS LESS THAN .010 POSITION TOL;
HOLE IS WITHIN TOLERANCE.

SIZE PRODUCED HOLE #2 = W/IN .305 - .310
HOLE ACTUAL LOCATION:

(HORIZONTAL X)	(VERTICAL Y)
ACTUAL - BASIC = X	ACTUAL - BASIC = Y
2.903 - 2.900 = .003	.803 - .800 = .003

FROM CHART: .003 X AND .003 Y = .0085
.0085 IS LESS THAN .010 POSITION TOL;
HOLE IS WITHIN TOLERANCE.

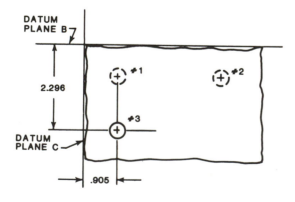

SIZE PRODUCED HOLE #3 = .308 (FIND ITS SIZE)
HOLE ACTUAL LOCATION:

(HORIZONTAL X)	(VERTICAL Y)
ACTUAL - BASIC = X	BASIC - ACTUAL = Y
.905 - .900 = .005	2.300 - 2.296 = .004

FROM CHART: .005 X AND .004 Y = .0128
ACTUAL SIZE OF HOLE #3 IS .308; THEREFORE, POSI-
TION TOL PERMISSIBLE USING MMC PRINCIPLES FOR
HOLE #3 IS .013 (.010 + .003 DEPARTURE FROM MMC);
.0128 IS LESS THAN .013; HOLE IS WITHIN TOLERANCE.

COMPARISON: ANSI Y14.5, ISO* SYMBOLS

CHARACTERISTIC	ANSI - Y14.5	ISO 1101
STRAIGHTNESS	—	—
FLATNESS	▱	▱
ANGULARITY	∠	∠
PERPENDICULARITY (SQUARENESS)	⊥	⊥
PARALLELISM	//	//
CONCENTRICITY	◎	◎
POSITION	⊕	⊕
CIRCULARITY (ROUNDNESS)	○	○
SYMMETRY	⊕	≡
PROFILE OF ANY LINE	⌒	⌒
PROFILE OF ANY SURFACE	⌓	⌓
RUNOUT (CIRCULAR)	↗	↗
RUNOUT (TOTAL)	↗↗	↗↗
CYLINDRICITY	⌭	⌭
DATUM FEATURE	- A -	A
MAXIMUM MATERIAL CONDITION (MMC)	Ⓜ	Ⓜ
REGARDLESS OF FEATURE SIZE (RFS)	Ⓢ	NONE (ASSUMED UNLESS SPECIFIED MMC)
LEAST MATERIAL CONDITION (LMC)	Ⓛ	NONE (PROPOSED)

*ISO - International Standards Organization

GLOSSARY

ACTUAL SIZE—An actual size is the measured size of the feature.

ANGULARITY—Angularity is the condition of a surface, axis, or center plane which is at a specified angle (other than 90°) from the datum plane or axis. Symbol: ∠

BASIC DIMENSION—A dimension specified on a drawing as BASIC (or abbreviated BSC) is a theoretical value used to describe the exact size, shape, or location of a feature. It is used as the basis from which permissible variations are established by tolerances on other dimensions or notes. A basic dimension is symbolized by boxing it: 1.270

BASIC SIZE—The basic size is that size from which limits of size are derived by the application of allowances and tolerances.

BILATERAL TOLERANCING—A bilateral tolerance is a tolerance in which variation is permitted in both directions from the specified dimension, e.g., .810 ± .010.

CENTER PLANE—Center plane is the middle or median plane of a feature.

CIRCULAR RUNOUT—Circular runout is the composite control of circular elements of a surface independently at any circular measuring position as the part is rotated through 360°. Symbol: ↗

CIRCULARITY—Circularity is the condition on a surface of revolution (cylinder, cone, sphere) where all points of the surface intersected by any plane (1) perpendicular to a common axis (cylinder, cone) or (2) passing through a common center (sphere) are equidistant from the center. Symbol: ○

CLEARANCE FIT—A clearance fit is one having limits of size so prescribed that a clearance always results when mating parts are assembled.

COAXIALITY—Coaxiality of features exists when two or more features have coincident axes, i.e., a feature axis and a datum feature axis.

CONCENTRICITY—Concentricity is a condition in which two or more features (cylinders, cones, spheres, hexagons, etc.) in any combination have a common axis. Symbol: ◎

CONTOUR TOLERANCING—See Profile of a Line or Surface.

CYLINDRICITY—Cylindricity is a condition of a surface of revolution in which all points of the surface are equidistant from a common axis. Symbol: ⌭

DATUM—A datum is a theoretically exact point, axis, or plane derived from the true geometric counterpart of a specified datum feature. A datum is the origin from which the location or geometric characteristics of features of a part are established.

DATUM AXIS—The datum axis is the theoretically exact center line of the datum cylinder as established by the extremities or contacting points of the actual datum feature cylindrical surface, or the axis formed at the intersection of two datum planes.

DATUM FEATURE—A datum feature is an actual feature of a part which is used to establish a datum.

DATUM FEATURE SYMBOL—The datum feature symbol contains the datum reference letter in a drawn rectangular box, e.g., $\boxed{\text{-A-}}$

DATUM LINE—A datum line is that which has length but no breadth or depth such as the intersection line of two planes, center line or axis of holes or cylinders, reference line for functional, tooling, or gaging purposes. A datum line is derived from the true geometric counterpart of a specified datum feature when applied in geometric tolerancing.

DATUM PLANE—A datum plane is a theoretically exact plane established by the extremities or contacting points of the datum feature (surface) with a simulated datum plane (surface plate or other checking device). A datum plane is derived from the true geometric counterpart of a specified datum feature when applied in geometric tolerancing.

DATUM POINT—A datum point is that which has position but no extent such as the apex of a pyramid or cone, center point of a sphere, or reference point on a surface for functional, tooling, or gaging purposes. A datum point is derived from a specified datum target on a part feature when applied in geometric tolerancing.

DATUM REFERENCE—A datum reference is a datum feature as specified on a drawing.

DATUM REFERENCE FRAME—A datum reference frame is a system of three mutually perpendicular datum planes or axes established from datum features on a basis for dimensions for design, manufacture and verification. It provides complete orientation for the features involved.

DATUM SURFACE—A datum surface or feature (hole, slot, diameter, etc.) refers to the actual part surface or feature coincidental with, relative to, and/or used to establish a datum.

DATUM TARGET—A datum target is a specified datum point, line, or area identified on the drawing with a datum target symbol ⊖ used to establish datum points, lines, planes, or areas for special function, or manufacturing and inspection repeatability.

DIMENSION—A dimension is a numerical value expressed in appropriate units of measure and indicated on a drawing.

FEATURE—Feature is the general term applied to a physical portion of a part, such as a surface, hole, pin, slot, tab, etc.

FEATURE CONTROL FRAME—The feature control frame is a rectangular box containing the geometric characteristic symbol and the form, orientation, profile, runout, or location tolerance. If necessary, datum references and modifiers applicable to the feature or the datums are also contained in the frame, e.g., $\boxed{\nearrow\ |\ .002\ |\ A}$

FEATURE OF SIZE—A feature of size is one cylindrical or spherical surface, or a set of two plane parallel surfaces, each of which is associated with a dimension. A feature such as a hole, shaft, pin, slot, etc. which has an axis, centerline or centerplane when related to geometric tolerances.

FIT—Fit is the general term used to signify the range of tightness or looseness which may result from the application of a specific combination of allowances and tolerances in the design of mating part features. Fits are of four general types: clearance, interference, transition, and line.

FLATNESS—Flatness is the condition of a surface having all elements in one plane. Symbol: ▱

FORM TOLERANCE—A form tolerance states how far an actual surface or feature is permitted to vary from the desired form implied by the drawing. Expressions of these tolerances refer to flatness, straightness, circularity, cylindricity. Symbols: ▱ — ○ ⌀

FULL INDICATOR MOVEMENT (FIR) (TIR) (FIM)—Full indicator reading is the total indicator movement reading observed with the dial indicator in contact with the part feature surface during one full revolution of the part about its datum axis. (Use of the international term FIM is recommended.)

Full indicator reading also refers to the full indicator reading observed while in traverse over a fixed noncircular shape.

GEOMETRIC CHARACTERISTICS—Geometric characteristics refer to the basic elements or building blocks which form the language of geometric dimensioning and tolerancing. Generally, the term refers to all the symbols used in form, orientation, profile, runout, and location tolerancing.

IMPLIED DATUM—An implied datum is an unspecified datum whose influence on the application is implied by the dimensional arrangement of the drawing; e.g., the primary dimensions are tied to an edge surface; this edge is implied as a datum surface and plane.

INTERFERENCE FIT—An interference fit is one having limits of size so prescribed that an interference always results when mating parts are assembled.

INTERRELATED DATUM REFERENCE FRAME—An interrelated datum reference frame is one which has one or more common datums with another datum reference frame.

LEAST MATERIAL CONDITION (LMC)—This term implies that condition of a part feature wherein it contains the least (minimum) amount of material, e.g., largest hole size and smallest shaft size. It is opposite to maximum material condition (MMC). Symbol used: Ⓛ

LIMIT DIMENSIONS (TOLERANCING)—In limit dimensioning only the maximum and minimum dimensions are specified. When used with dimension lines, the maximum value is placed above the minimum value, e.g. $\frac{.300}{.295}$. When used with leader or note on a single line, the minimum limit is placed first, e.g., .295 .300.

LIMITS OF SIZE—The limits of size are the specified maximum and minimum sizes of a feature.

LINE FIT—A line fit is one having limits of size so prescribed that surface contact or clearance may result when mating parts are assembled.

LOCATION TOLERANCE—A location tolerance states how far an actual feature may vary from the perfect location implied by the drawing as related to datums or other features. Expressions of these tolerances refer to the category of geometric characteristics containing position and concentricity (formerly also symmetry). Symbols: ⊕ ◎

MAXIMUM DIMENSION—A maximum dimension represents the acceptable upper limit. The upper limit may be considered any value greater than the minimum specified.

MAXIMUM MATERIAL CONDITION (MMC)—Maximum material condition is that condition where a feature of size contains the maximum amount of material within the stated limits of size, e.g., minimum hole diameter and maximum shaft diameter. Symbol used: Ⓜ

MINIMUM MATERIAL CONDITION— See Least Material Condition.

MODIFIER (MATERIAL CONDITION SYMBOL)—A modifier is the term sometimes used to describe the application of the "maximum material condition," "regardless of feature size" or "least material condition" principles. The modifiers are maximum material condition (MMC), symbol Ⓜ ; regardless of feature size (RFS), symbol Ⓢ ; and least material condition (LMC), symbol Ⓛ .

MULTIPLE DATUM REFERENCE FRAMES—Multiple datum reference frames are more than one datum reference frame on one part.

NOMINAL SIZE—The nominal size is the stated designation which is used for the purpose of general identification, e.g., 1.400, .060, etc.

NORMALITY—See Perpendicularity.

ORIENTATION TOLERANCE—Orientation tolerances are applicable to related features, where one feature is selected as a datum feature and the other related to it. Orientation tolerances are perpendicularity, angularity and parallelism. Symbols: ⊥ ∠ ∥

PARALLELEPIPED—Shape of tolerance zone. The term is used where total width is required and to describe geometrically a square or rectangular prism, or a solid with six faces, each of which is a parallelogram.

PARALLELISM—Parallelism is the condition of a surface, line, or axis which is equidistant at all points from a datum plane or a datum axis. Symbol: ∥

POSITION TOLERANCE—A position tolerance (formerly called true position tolerance) defines a zone within which the axis or center plane of a feature is permitted to vary from true (theoretically exact) position. Symbol: ⊕

PROFILE OF A LINE—Profile of a line is the condition permitting a uniform amount of profile variation, either unilaterally or bilaterally, along a line element of a feature. Symbol: ⌒

PROFILE OF A SURFACE—Profile of a surface is the condition permitting a uniform amount of profile variation, either unilaterally or bilaterally, on a surface. Symbol: ⌓

PROFILE TOLERANCE—Profile tolerance controls the outline or shape of a part as a total surface or at planes through a part. Symbols: ⌓ ⌒

PROJECTED TOLERANCE ZONE—A projected tolerance zone is a tolerance zone applied to a hole in which a pin, stud, screw, or bolt, etc. is to be inserted. It controls the perpendicularity of the hole to the extent of the projection from the hole and as it relates to the mating part clearance. The projected tolerance zone extends above the surface of the part to the functional length of the pin, screw, etc., relative to its assembly with the mating part. Symbol: Ⓟ

REGARDLESS OF FEATURE SIZE (RFS)—This is the condition where the tolerance of form, runout, or location must be met irrespective of where the feature lies within its size tolerance. Symbol: Ⓢ

ROUNDNESS—See Circularity.

RUNOUT— Runout is the composite deviation from the desired form of a part surface of revolution during rotation (360°) of the part on a datum axis. Runout tolerance may be "circular" or "total." Symbols:

RUNOUT TOLERANCE—Runout tolerance states how far an actual surface or feature is permitted to deviate from the desired form implied by the drawing during full rotation of the part on a datum axis. There are two types of runout: circular runout and total runout. Symbols:

SIZE TOLERANCE—The size tolerance states how far individual features may vary from the desired size. Size tolerances are specified with either unilateral, bilateral, or limit tolerancing methods.

SPECIFIED DATUM—A specified datum is a surface or feature identified with a datum feature symbol. (See Datum Feature Symbol.)

SQUARENESS—See Perpendicularity.

STRAIGHTNESS—Straightness is a condition where an element of a surface or an axis is a straight line. Symbol: ▬▬

SYMMETRY—Symmetry is a condition in which a feature (or features) is symmetrically disposed about the center plane of a datum feature. Symbol: (Symbol ≡ replaced by ⊕ in latest 1982 standards).

TOLERANCE—A tolerance is the total amount by which a specific dimension may vary, thus, the tolerance is the difference between limits.

TOTAL INDICATOR READING (TIR) (FIR) (FIM)—Total indicator reading is the full indicator reading observed with the dial indicator in contact with the part feature surface during one full revolution of the part about its datum axis. Total indicator reading also refers to the total indicator reading observed while in traverse over a fixed noncircular shape. (Use of the international term FIM is recommended.)

TOTAL RUNOUT—Total runout is the simultaneous composite control of all elements of a surface at all circular and profile measuring positions as the part is rotated through 360°. Symbol:

TRANSITION FIT—A transition fit is one having limits of size so prescribed that either a clearance or an interference may result when mating parts are assembled.

TRUE POSITION—True position is a term used to describe the perfect (exact) location of a point, line, or plane of a feature in relationship with a datum reference or other feature.

UNILATERAL TOLERANCE—A unilateral tolerance is a tolerance in which variation is permitted only in one direction from the specified dimension, e.g., $1.400 \pm {.000 \atop .005}$

VIRTUAL CONDITION—Virtual condition of a feature is the collective effect of size, form, and location error that must be considered in determining the fit or clearance between mating parts or features. It is a derived size generated from the profile variations permitted by the specified tolerances. It represents the most extreme condition of assembly at MMC.

INDEX